PRINTED IN GREAT BRITAIN

PREFACE

Every year for more than a century, the Royal Institution has invited some man of science to deliver a course of lectures at Christmastide in a style "adapted to a juvenile auditory". In practice, this rather quaint phrase means that the lecturer will be confronted with an eager and critical audience, ranging in respect of age from under eight to over eighty, and in respect of scientific knowledge from the aforesaid child under eight to staid professors of science and venerable Fellows of the Royal Society, each of whom will expect the lecturer to say something that will interest him.

The present book contains the substance of what I said when I was honoured with an invitation of this kind for the Christmas season 1933–4, fortified in places with what I have said on other slightly more serious occasions, both at the Royal Institution and elsewhere.

It is a pleasure to acknowledge many courtesies and return thanks for much valuable help. I am indebted to Sir T. L. Heath for permission to borrow largely from his *Greek Astronomy* and other books; to many Institutions, Publishers and private individuals for the loan of negatives, prints, blocks, etc., and permission to reproduce these in my book—detailed acknowledgment is made in the List of Illustrations. Finally, I have to thank Sir Thomas Heath and my sister, Gertrude Jeans, for help in reading the proofs, and the staff of the Cambridge University Press for their usual careful help in producing the book.

J. H. JEANS

Dorking
August 1934

CONTENTS

LIST OF ILLUSTRATIONS

List of Illustrations

List of Illustrations

CHAPTER I

THE EARTH

These are restless days in which everyone travels who can. The more fortunate of us may have travelled outside Europe to other continents—perhaps even round the world—and seen strange sights and scenery on our travels. And now we are starting out to take the longest journey in the whole universe. We shall travel—or pretend to travel—so far through space that our earth will look like less than the tiniest of motes in a sunbeam, and so far through time that the whole of human history will shrink to a tick of the clock, and a man's whole life to something less than the twinkling of an eye.

As we travel through space, we shall try to draw a picture of the universe as it now is—vast spaces of unthinkable extent and terrifying desolation, redeemed from utter emptiness only at rare intervals by small particles of cold lifeless matter, and at still rarer intervals by those vivid balls of flaming gas we call stars. Most of these stars are solitary wanderers through space, although here and there we may perhaps find a star giving warmth and light to a family of encircling planets. Yet few of these are at all likely to resemble our own earth; the majority will be so different that we shall hardly be able to describe their scenery, or imagine their physical condition.

As we travel through time, we shall try to extend this momentary picture into a sort of cinematograph film that will shew us not only the present, but also the past and the future, of the universe. We shall see the sky as it was a million years ago, a thousand million, and possibly even a million million years ago; we shall watch vast colonies of stars, each like the sands of the seashore in number, being born, living their lives, and finally dying. As one tiny incident in the great drama, we shall watch one inconspicuous grain of sand—our sun—being broken up in great turmoil and finally producing a family of planets. We shall watch one of the smaller of these planets—our earth—coming into being as a globe of hot gas which gradually cools, and ultimately becomes a suitable abode for life. In due course we shall see life appearing, and finally man arriving, taking possession of his tiny speck of dust in space, surveying with astonishment the strange universe in which his life is cast, and looking wonderingly and perhaps anxiously and fearfully into the future.

Before we start on our long journey, let us pause to examine our own home in space—the earth. We shall learn a lot from it that will be useful in our travels. We know that it is globular in shape; we discover this by travelling over it and mapping it out, by watching ships coming over the horizon, or by examining the shape of its shadow when this passes over the face of the moon at an eclipse. It may sound a simple matter to do all this, but the human race had inhabited the earth for hundreds of thousands of years before doing it. For until the last few hundreds of years most people thought the earth was flat, and a few misguided people still think it is. The ancient Greeks, including Homer, thought the earth was a flat circular disc, with Oceanus, the ocean—which they regarded as a river—flowing all around it. The dome of heaven covered this much as a dish-cover covers a dish. Probably the Greek Pythagoras, who was born about 570 B.C., was the first to maintain that the earth had a globular shape.

We also know that the earth is rotating. Day after day and night after night, we see sun, moon and stars rising in the east, moving in stately procession across the sky, and sinking in the west; and ever since the dawn of human intelligence men must have noticed the same thing. But so long as they thought of the earth as a flat plain, it was easier to picture the dome of heaven as turning over the earth than to imagine that the earth might be turning under the dome of heaven. Even Pythagoras, who believed that the earth was a globe floating in space, did not suspect that it turned round under the stars. He imagined that it stood at rest at the centre of the universe, and that the stars were attached to a sphere which turned around it from east to west. So far as we know, Heraclides of Pontus (about 388–315 B.C.) was the first to state perfectly clearly that it was the earth itself which turned round, and that this was why the heavenly bodies appeared to move across the sky.

It is not difficult to prove for ourselves that it is we who are moving round under the stars, and not the stars that are moving round above our heads. Now that we all drive cars, we are all familiar with the property of matter that we describe as "inertia". About a century after Christ, Plutarch explained it in the words "Everything is carried along by the motion natural to it, if it is not deflected by something else". Fifteen hundred years later, Isaac Newton described the same property of matter by saying that every body perseveres in its state of rest, or of uniform motion in a straight line, unless it is compelled to change by forces impressed on it. When our car is running freely, stopping the engine does not stop the car; the momentum of the car still carries it forward, and to stop it we must either put on the brakes, or wait until friction and air-resistance brake the motion in a more leisurely manner. Not only every object, but every part of an object, seems to want to continue its present motion, and will only make a change if something pulls on it and compels it to do so. If we turn the steering-wheel of our car, we can make the lower part of the car follow the front wheels, but the upper part will seem to want to continue on its old course; if we turn the wheel too abruptly, there is a danger, as we know, that the car will overturn. Or, if the road is icy or muddy, so that the wheels get no grip on the road, the whole back part of the car will tend to follow its old course, so that the car may skid. We shall encounter this property of inertia very often on our journey through time and space.

It is important to us at the moment because it provides us with the simplest and most convincing proof that the earth actually is rotating. If we swing a heavy ball or weight, pendulum-wise, at the end of a string, we shall find that it keeps on swinging in the same direction in space, no matter how much the top of the string is twisted or turned about; we can no more steer the swing of the pendulum in space by turning the top of the string than we can steer a car on ice by turning the steering-wheel.

Now let us set our pendulum swinging in such a direction that it swings towards and away from some clearly defined landmark, such as a church tower. As we want the motion to continue for a long time, we had better take a really heavy weight and suspend it from a high roof; if we try the experiment on a less massive scale, the pendulum will be stopped too soon by air resistance.

If the earth were standing still in space, our pendulum would naturally continue swinging towards and away from the tower, until the resistance of the air brought it to rest. Instead of this, we shall find the direction of its swing moving gradually farther and farther away from the church tower. The true direction of the swing of the pendulum cannot have changed, so we can only conclude that the church tower must have moved. And this, indeed, is what has happened; the rotation of the earth has carried it round.

Now let us start on our travels by going to the North Pole, and let us take our pendulum with us and perform our experiment there. If we disregard the earth, and keep our eyes fixed on the sky, we shall see that the swinging pendulum moves towards and away from the same stars throughout its whole motion; if, for instance, we start it swinging towards Arcturus, it will keep on swinging towards and away from Arcturus. This proves that Arcturus stays always in the same direction in space. If we now look down to the earth, we shall be able to watch the earth's surface turning round under our non-turning pendulum at the rate of a revolution once every 24 hours—or, to be more precise, every 23 hours 56 minutes and 4·1 seconds. In other latitudes the result of the experiment is less easy both to describe and to explain.

This experiment is generally known as Foucault's experiment. The French physicist Foucault performed it in public in 1852, suspending his pendulum from the dome of the Panthéon in Paris. Thousands of people watched, and, as they saw the pendulum change its direction relative to the walls of the Panthéon, many averred that they could feel the earth turning under their feet.

The same principle of inertia provides a second, but rather less direct, proof of the earth's rotation. We who live in England are so accustomed to the incessant and rapid changes in our own weather that we almost forget that there are large stretches of the earth over which the weather hardly varies at all. It is always hot in the vicinity of the equator, and as winds drag air over these hot regions, the air itself becomes heated and tends to rise upwards, like the hot air in a stuffy room or the hot gases in a chimney. In the same way, when the winds drag air across the Arctic and Antarctic regions, this air becomes cooled and so tends to fall earthwards.

If the earth were not rotating at all, this local heating and cooling of air would set the whole atmosphere into a state of steady circulation in a north-south direction. Air would descend at both poles; the pressure of other air descending behind it would then push it along the earth's surface towards the equator, where it would rise upwards and move back to the poles through the upper reaches of the atmosphere. Such a circulation actually occurs, but is almost concealed by other and more complicated motions produced by the rotation of the earth.

The rotating earth drags the whole circulating system of air round with it, but the latter can never quite keep pace with the solid earth which is forcing its motion. A mountain or other point on the earth's surface in Norway is moving round the axis of the earth at about 500 miles an hour, while one near the equator is moving at about 1000 miles an hour. Now the frictional drag of the earth is never quite forcible enough to speed the air up from 500 to 1000 miles an hour in the course of its southerly journey from Norway to the equator. The earth's mountains and surface are not rough and spiky enough to get a perfectly firm grip on the air, so that this is always slipping backwards a bit—as a motor-car does when the clutch is not holding perfectly. When we feel the air slipping back in this way, we say there is a wind blowing from the east to the west.

This is the origin of the trade-winds which blow steadily westward on both sides of the equator. If the earth were not rotating there would be nothing to cause the trade-winds, so that we can think of these winds as providing a proof of the earth's rotation. It is easier to sail westward than eastward, because in sailing westward the inertia of the air around us keeps us from participating in the full motion of the earth. In sailing eastward, we have the more serious task of overtaking the earth in its motion.

Shortly after Heraclides had explained the rotation of the earth, Eratosthenes of Alexandria measured the earth's size with great skill and surprising success. He believed, with most people of his time, that the sun's distance was enormously great in comparison with the dimensions of the earth. If, then, the earth had been completely flat, the sun would have been directly overhead at all places at the same time. Actually he found that when it was overhead at Syene (the modern Assouan), it was not overhead at Alexandria, which lay 5000 stades to the north. As the sun's rays could not come from different directions at the two places, he argued that the "overhead" directions must be different. Actually he found they were different by a fiftieth of a circle, or $7\frac{1}{5}$ degrees—when the sun was directly overhead at Syene, it was $7\frac{1}{5}$ degrees from the zenith at Alexandria. Hence he concluded that the earth's surface curved through $7\frac{1}{5}$ degrees between the two places, or, as we should say to-day, that the difference of latitude between the two places was $7\frac{1}{5}$ degrees.* An easy calculation shewed that the circumference of the whole circle of the earth must be fifty times 5000 stades, or 250,000 stades. Eratosthenes subsequently amended this to 252,000 stades, which was probably equivalent to about 24,662 of our English miles. As the actual circumference of the earth measured in a north-south direction is 24,819 miles, while that around the equator is 24,902 miles, we see that Eratosthenes' measurement was in error by less than one per cent.

Let us take yet another illustration of the principle of inertia,

* The true difference is about $7° 7'$.

which tells us that objects continue moving in a straight line unless something pulls them away from it. We know that if we are swinging a weight round at the end of a string and the string suddenly breaks, the weight will immediately fly off at a tangent into space. Now that the string is broken, the inertia of the

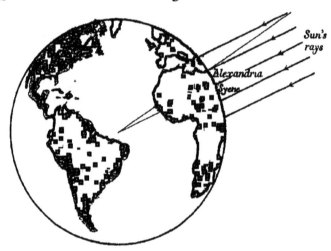

Fig. 1. Eratosthenes found that when the sun's rays fell vertically at Syene, they were a fiftieth of a whole circle away from the vertical at Alexandria. He concluded that the circumference of the earth was fifty times the distance from Alexandria to Syene.

weight carries it on in a straight line; before the string broke something must have been pulling on the weight to keep it moving in a circle; this was of course the pull of the string.

Now objects at the earth's equator are in a similar position to the weight at the end of the string. The earth's rotation carries them round and round in a circle 24,902 miles in circumference once every 24 hours, so that their speed is rather more than

1000 miles an hour. The principle of inertia tells us that they would continue their motion in a straight line, and so fly off at a tangent into space, were it not that something is continually stopping them from doing so—pulling them out of the straight line in which they would otherwise move.

We describe this something as the "gravitational pull" of the earth. It pulls on our bodies with such force that we find ourselves unable to jump more than a few feet into the air, and of course it must exert a correspondingly powerful pull on other objects. Yet it is not all-powerful. The faster a body moves, the greater the pull needed to keep it to a circular path—as we discover if we return to our weight and swing it round faster and ever faster at the end of our string. The earth's pull can easily hold down objects moving at 1000 miles an hour, but with objects moving faster there is less margin to spare. The margin would disappear entirely if the earth were suddenly to start spinning at seventeen times its present speed, so that we had an 85-minute day. We should then see the surprising spectacle of all the objects at and near the earth's equator rising from the ground and flying off at tangents into space, the air and the sea of course accompanying them on their journey. Objects reposing on the earth's surface are rather like drops of rain on the surface of a bicycle wheel: so long as the wheel spins slowly, nothing happens, but when it spins fast they fly off and do not come back.

With things as they are, objects at the equator are very far from being thrown off into space, yet they shew a certain tendency in that direction. For instance, a man at the equator is able to jump to a height of 6 feet with less physical effort than anywhere else on the earth's surface, because his speed of 1000 miles an hour helps him to counteract the earth's gravitational pull. For this reason, athletic records made in different latitudes are not strictly comparable; there ought to be a handicap for nearness to the equator.

We can see further evidence of the same tendency in the fact that the earth itself bulges out at the equator. This is often described by saying that the earth is flattened like an orange, but in actual fact its longest diameter is only 27 miles longer than the shortest—a difference of only about 1 part in 300—and an orange with no more flattening than this would look perfectly round to casual observation. Yet although the earth's flattening is almost inappreciable we shall soon come to planets which are spinning round so fast that their flattening is obvious at the merest glance, while later in our travels we shall come upon bodies of other kinds which are spinning so fast that objects are actually flying off their equators into space.

Our earth is not only like an orange in being flattened, but also in having a rough skin or peel covered with elevations—its mountains and valleys. But again this comparison exaggerates the earth's irregularities. It would only give us a fair picture, properly drawn to scale, if the earth were studded with mountains 50 miles in height; whereas actually the world's highest mountain, Everest, is less than 5½ miles high. On a 12-inch geographical globe even the overlap of the paper represents a precipice about 7 miles high, and so represents greater irregularities than occur on earth. Taking it all in all, the earth comes very near indeed to being a perfect sphere, and certainly resembles a billiard ball much more than an orange.

The comparison with an orange fails in yet a third respect. The earth's mountains do not occur in regular formation, like the excrescences on the peel of an orange, but in irregular ridges and ranges—more like the foldings in the peel of a shrunken apple. And this last happens to be a very good and useful comparison, because the earth's mountain ridges actually are due to shrinkage; they exist for precisely the same reason as the foldings in the apple-peel.

I am afraid we shall not fully understand this, until we have explored quite a bit, both in space and in time—far back into the earth's past history, and far down into its interior.

How are we to explore inside the earth? We might, of course, dig a hole, as the mining engineer does in his search for coal, or bore down, as the oil engineer does in his search for oil; but neither of these methods will take us any appreciable way towards the centre of the earth. Oil-borings only go about 8000 feet down, and coal mines only half as far; the deepest of man-made holes are only the tiniest of pin-pricks in the skin of our apple, and take us nowhere near the central parts.

Because of this, it is surprising but true that until quite recently we knew more about the state of the most distant stars than about the state of the earth a few miles under our feet. But the new science of seismology has shewn us how to probe thousands of times deeper than any hole can take us—and indeed to the very centre of the earth.

There are many indications that the pressure in the earth's interior is for ever varying, and that the earth's structure is for ever gradually shifting and changing as it yields to these ever-varying pressures. Occasionally this gradual yielding gives place to a sudden snap or break which jars the whole earth—an earthquake.

(When an earthquake occurs, waves start out from the point of breakage and travel in all directions through the whole earth—just as, when we throw a stone in a pond, waves start out from the point of impact and travel over the whole pond. When these waves finally emerge at the surface of the earth, they bring with them a whole fund of information as to the conditions they have encountered on their long journey through the earth's interior. Consequently, these waves are recorded and studied at hundreds of observatories, which are well sprinkled over the whole surface of the earth. These obtain records of hundreds of earthquakes every year, most of which, happily, are too slight to do any damage to life or property, and would escape observation entirely were it not for the extreme sensitiveness of the instrument known as the seismograph, which is used to detect them.)

The essentials of such an instrument are shewn diagrammatically in fig. 3 (facing p. 20). It consists primarily of a long arm or horizontal pendulum, swinging freely on a vertical pivot which is in some way connected with the rock or soil of the solid earth. When the earth is shaken, a wave comes along which gives a jar to the pivot and so causes the boom to start swinging; a pen connected with the end of the boom automatically records the motion on a strip of moving paper. It is necessary to use two such instruments simultaneously, one boom pointing in a north-south direction, and the other east-west. If only one boom were in use, it would give no record of earthquake waves travelling in the direction in which it was pointing.

If the instrument is to serve its purpose, the boom must be suspended very delicately, and then, unfortunately, it cannot be restrained from recording all sorts of jarrings of the earth, what-ever their cause. For instance, it faithfully records the passage of every train, omnibus, or motor-lorry, so that unless the observer wishes to be continually distracted by all these, he had better install his seismograph in a quiet place. Even then he will find that the pounding of the sea on our coasts shakes our whole island, and his seismograph with it, so that from observations taken far inland, he can tell whether it is rough or fine out at sea. The records obtained at the Indian Observatory of Colaba are found to vary quite definitely in quality with the conditions in the Bay of Bengal and the Arabian Sea. Storms as much as 1000 miles distant have been detected in this way, and attempts have been made to predict the approach of cyclones or mon-soons.

The experienced observer finds no difficulty in distinguishing between local shocks such as these, which affect only a small part of the earth's surface, and true earthquakes, which affect the whole earth. A portion of an actual seismograph record is shewn in fig. 4 (facing p. 20). The big open waves at the right-hand of the second line down are the record of true earthquake waves; all the smaller waves represent minor earth tremors, produced by untraced causes.

When their seismographs record a true earthquake, the various observatories note the times at which the shocks reach them, and from the differences between these, they can deduce the speed with which the waves have travelled through the solid earth.

If the earth's interior were uniform in structure and composition, earthquake waves would always travel at the same uniform speed. Actually, however, it is found that waves which have been deep down into the earth's interior travel at a far higher average speed than those which have stayed near the surface. On the other hand, waves which have been to the same depth always travel at the same average speed. This is true whether their path has been in a north-south, or east-west, or any other direction, also whether they have travelled under a continent or under an ocean, under the old world or under the new. This shews that the earth is everywhere of similar substance and composition at the same depth, although it may be different at different depths.

Thus we may think of the earth's interior as a series of spherical layers surrounding one another like the skins of an onion, or again we may compare it to a globular parcel wrapped up in a great number of wrappings.

When an earthquake occurs, the waves which are most noticeable, and also do most damage, are the "surface" waves which travel along the ground. In addition to these, there are two distinct kinds of waves which travel through the earth's interior —the "primary" waves which consist of longitudinal motions, and the "secondary" waves which consist of transverse motions. Neither a liquid nor a gas can transmit transverse waves, so that the secondary waves can only travel through solid matter. Actually they are found to travel through the whole interior of the earth, except for a region, extending for about 2200 miles in every direction from the centre of the earth, which is known as the "central core" of the earth. It seems safe to conclude that the whole interior of the earth is solid, except for the central core. This may be either liquid or gaseous, unless indeed it consists of matter in some state of which we have no experience. It seems likely that it consists of very heavy liquid, perhaps ten or twelve times as dense as water. This may be mainly molten iron, perhaps mixed with nickel, and possibly, as geologists have thought, similar in its chemical composition to the iron meteorites which often fall on the earth's surface. It is true that these substances are not normally ten or twelve times as dense as water, but we do not normally meet them under high pressures, and the pressure inside the central core must be im-mense, since it has to support the weight of the greater part of the earth. A rough computation indicates that the pressure at the surface of this core may be about 7500 tons to the square inch, which is a million times the pressure of the atmosphere at the earth's surface. The pressure at the centre of the earth will be even greater, perhaps about 10,000 tons to the square inch.

We may think of the central core as our parcel. Immediately outside it comes the first wrapping—a layer of matter about 1700 miles thick, which is sometimes described as the "barysphere". This transmits both kinds of earthquake waves at speeds which shew that it consists of heavy solid matter, more rigid than steel. Even inside the barysphere the waves do not travel at a uniform speed; as with the earth as a whole, those waves which go deepest travel fastest, shewing that the deeper layers are more rigid than the upper. Inside the barysphere there may perhaps be a gradual transition from such heavy substances as iron and nickel in the lowest layers to those lighter substances of which the earth's surface rocks are formed.

The barysphere extends to within about 50 miles of the earth's surface, so that the remaining wrappings are comparatively thin. These are believed to consist of rocky substances, and are often described collectively as the "lithosphere", or sphere of rocks. Seismologists detect three distinct layers, through which earthquake waves travel in different ways and at different speeds, giving some indication at least as to the structure of the rocky layers. There is no general agreement about the lowest layer of all, but it is thought that the middle layer is probably basaltic, while the uppermost is almost certainly granitic.

Fig. 2 (p. 18) represents the arrangement of the earth's interior as indicated by the evidence of seismology.

The inner core and the various wrappings we have so far described constitute the essential and permanent body of the earth. If we continue to compare the earth to an apple, the central core is its core, the barysphere its flesh, and the lithosphere its skin; such a picture will not be too badly out of drawing in the matter of relative proportions. But outside all these come other layers and wrappings of a more accidental, ephemeral and variable nature, which we may perhaps compare to layers of dust and drops of rain on the skin of our apple.

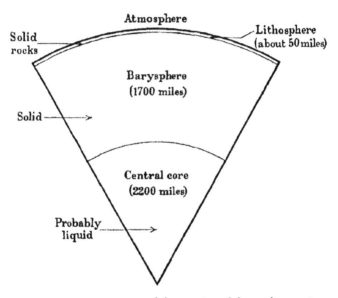

Fig. 2. Diagrammatic representation of the interior of the earth as conjectured from observations on earthquake waves. The height of the highest mountains is only a tenth of the thickness of the lithosphere, and so is less than the thickness of the printed line which represents the surface of the earth.

First of all, like the dust on the apple, come the layers known as "sedimentary", which we shall describe more fully in a moment. There are several such layers, and their total thickness—the thickness of the layer of dust on our apple—varies from many miles to almost nothing, since places can be found (as in fig. 11, p. 30) where the granite rocks of the lithosphere come up almost to the surface.

Next, like drops of rain on the apple, we come to the layer of water which we describe as the ocean; the thickness of this layer varies from 5 miles in the great ocean deeps to nothing at all at places where the dry land emerges from the sea.

Finally, outside all, comes the atmosphere, consisting of two layers called the "troposphere" and the "stratosphere", which we shall discuss in detail in the next chapter.

We see that the earth consists of a great number of distinct shells of matter. In a general way, the inner shells are found to consist of heavier matter than the outer, as though the heavier substances had sunk deep down into the body of the earth, while the lighter had floated to the top. But the separation is far from complete, and some of the heaviest known substances, such as lead, quicksilver and gold, are found in the outermost crust of the earth.

We shall see later how the earth probably started its existence as a hot mass of gas. It was born in a cataclysm which would probably stir up its various constituent substances, and mix them up, even if they were not thoroughly mixed already. Then as peace and calm succeeded to turmoil and confusion, the lighter substances would begin to float upwards, while the heavier would sink towards the centre of the earth.

All this time the earth is cooling; at last it begins to liquefy, and after this to solidify. When a piece of the earth has once solidified, its various constituent substances are no longer able either to sink or rise; they are trapped in the solid mass, and must stay for ever where they were when the process of petrifaction overtook them. The distribution of light and heavy substances in the earth's crust and interior shews that the process of arrangement was well advanced, but not complete, when the earth solidified.

The outermost layers of the earth, having no blankets around them to keep the heat in, would of course cool most rapidly, and so would be the first to solidify. When this had happened, the earth would consist of a solid outer crust enclosing a hotter interior of gas and liquid—rather like a mince-pie, in which we know that a deceptively cool exterior often conceals an interior which is too hot to eat. Just as with a mince-pie left to itself, this would be succeeded by a stage in which the inner layers would also begin to cool, and probably also to shrink, since most substances, and especially gases, shrink as they cool.

The crust of an ordinary mince-pie can easily support its own weight, but the crust of a mince-pie which weighed a million tons would not be able to do so, and it must have been the same with the far more massive crust of the earth. As the inner layers shrank away from under it, and no longer supported its weight, it must gradually have caved in upon these inner layers to find support. In so doing, it was faced with the problem of how suddenly to grow smaller although it had already ceased contracting—a problem which it solved in the only possible way, by crumpling up into wrinkles and folds, just as an apple does when its softer centre shrinks with the onset of old age. Fig. 5 on

PLATE I

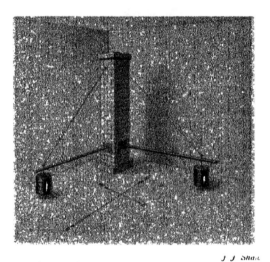

J J Shaw

Fig. 3. The essential parts of a seismograph Any vibration in the earth is transmitted to the brick pile, and sets the two booms swinging through minute angles These swings are amplified by mechanical levers and recorded as in fig 4 below

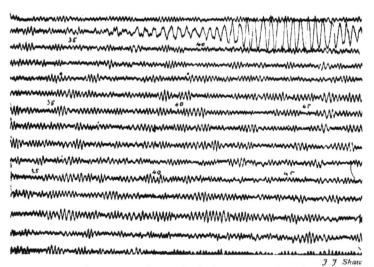

J J Shaw

Fig 4 A portion of a seismograph record The big waves to the right in the second line down were caused by a true earthquake of considerable violence All the others represent mere tremors such as might be produced by wind, sea-waves or traffic. The numbers indicate minutes.

PLATE II

Geological Museum

Fig 5   Layers of black slate and limestone at the south-east end of the island of
Kerrera, off Oban, shewing crumpling and folding The geological hammer on the
left shews the scale, but the same thing occurs on a scale 1000 times as big, and also
on a scale 1000 times as small

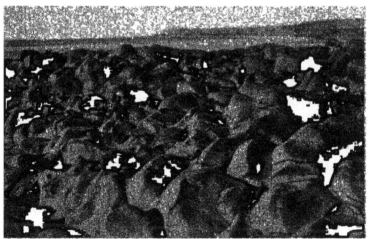

Geological Museum

Fig 6   Lava south of Ballantrae on the Ayrshire coast This flowed down into the
sea, perhaps 400 million years ago, and immediately solidified into the shapes it has
retained from that day to this.

Plate II shews a nearly vertical section of the earth, exposing the crumpling and folding of layers of slate and limestone; the layers of primitive rock must have crumpled in much the same way.

In some such way as this the earth formed its mountain ridges and valleys. The process is not entirely ended yet; the earth's surface is still moving slightly, falling in here and being pushed up there, so that new elevations and depressions are for ever being formed. Occasionally a sudden slip may result in an earthquake, such as we have already discussed. At other times the steady pressure of the falling or fallen crust may squeeze the hot material up through cracks and crevices until it emerges on the surface of the earth, as we see in volcanoes, oil wells, and in the spoutings of hot water known as geysers and hot springs; such happenings must have proceeded with incomparably greater vigour in the early days of the earth's history, and have left their marks very unmistakably on its present condition.

For, although there are few active volcanoes on the earth now, the number of mountains which shew evidence of having once been volcanic is enormous. Immense streams of lava and molten rock which they poured out in long-past ages still lie spread over large parts of the surface of the earth, and form the layers of rock which we describe as "igneous"—rocks laid down by fire. Fig. 6 shews a lava flow at Ballantrae, on the Ayrshire coast, which must have flowed directly down into the sea, immediately became petrified into its present "pillow" formation, and has retained its original form, through perhaps 400 million years, to the present day. The basaltic rocks of the well-known "Giant's Causeway" in Antrim form evidence of a similar outpouring of molten rock, which must have crystallised at once into its present hexagonal form. These rocky outpourings from primaeval volcanoes provide us with true samples of the substance of the earth's interior. Water and gases must have been forced up in a similar manner, and would of course make their contribution to the earth's ocean and atmosphere.

When the earth's crust fell in upon the shrinking inner mass, its wrinkles would not form entirely at random. For the crust is not likely to have been absolutely uniform in structure; it probably consisted of lighter and heavier parts. On the whole, the lighter parts would most readily be thrust up to form mountain ranges, while the heavier parts would tend to sink to the bottom of the folds, and form valleys and sea-bottoms. Thus we should naturally expect the mountains to be of lighter substance than the bottom of the sea—of fewer tons to the cubic yard. Recent careful measurements have shewn that this is actually the case.

The scientist does not try to sample his mountains and sea-bottoms by taking out a cubic yard here and there; such a method would be too crude for the mountains, and impossible for the sea-bottoms. In the past he used to take a long pendulum—rather like that from a grandfather clock, but made with the utmost scientific precision—to the top of a mountain, and try to discover the structure of the mountain from the behaviour of the pendulum. In recent years, the pendulum has been replaced by a more intricate instrument, but the general principle under-lying its use is much the same.

The top of a mountain is farther than the plain below from the centre of the earth, with the result that the earth's gravita-tional pull is less forcible there than down below. Thus, when a pendulum is pulled aside and set swinging up there, its bob moves more slowly to its lowest position, and so takes a longer time to reach this position than it would on the plain below. In other words, a pendulum which is taken to the top of a mountain will begin to lose time. We can calculate exactly how much time it would lose if the mountain were composed of the average stuff of the earth's crust. Always it loses a little more than this, shewing that the mountain is of lighter substance than the average. When the pendulum is taken to the bottom of the sea in a submarine we have exactly the reverse situation; it gains more than it would if the sea-bottom were of average substance, shewing that the sea-bottom is of heavier substance than the average.

Recently, a theory known as isostasy has given greater preci-sion to all of these ideas. It asserts, in brief, that mountains stand up above the level of the land for just the same reason that ships stand up above the level of the sea—because they float. It also supposes that, as with a ship, their total weight determines the height at which they float. A ship whose total weight—hull, cargo, crew, captain, and all—is 30,000 tons will float at a height at which she displaces exactly 30,000 tons of water; in other words, if she were suddenly lifted out of the water, she would leave a hole which it would take 30,000 tons of water to fill.

This, of course, is in accordance with the principle that Archimedes discovered 2200 years ago.

The theory of isostasy supposes that the height at which mountains float is determined in precisely the same way. The mountains are not of course supposed to be floating in water, or in any true liquid, but in some inner layer of the earth's substance which is plastic enough to behave like a liquid. Ordinary pitch, as we know, looks like a solid, but will yield to long-continued pressure just as a liquid does to momentary pressure. Pitch is plastic enough to yield in a matter of a few hours or days, ice in months or years (as we see in the flow of glaciers), and glass in years or centuries. The substance we are now considering will serve its purpose if it yields in millions of years. Various calculations suggest that we have to go to a depth of perhaps 20 miles to reach this plastic layer. Now it is a matter of common experience that pitch and other substances become more plastic— i.e. flow more easily—when they are heated; the same is probably true of the substance of the earth's crust, so that the heat at a depth of 20 miles or so may well provide the small degree of plasticity needed. The theory tells us that a mountain which weighs a million million tons will float at just that height at which it displaces a million million tons of this plastic inner layer. The most refined and careful measurements of which science is capable indicate that this theory gives an accurate account of the observed heights of the mountains.

I must here digress to tell you of a more recent theory, put forward by a German scientist, Wegener, which is perhaps even more interesting, although it has not yet gained such widespread acceptance from scientists. According to this theory, the continents and larger islands also are floating, not only like ships, but like independent ships which can approach towards and recede from one another. The old and new worlds are supposed originally to have formed a single big ship, which suffered shipwreck and broke into two, after which the parts drifted away from one another, the one forming Africa and Europe, while the other formed the American continent. It is claimed as evidence that if the New World were towed about 3150 miles to the E.N.E. it would fit very prettily on to the Old World, the point of Brazil on which Pernambuco stands fitting into the Bay of the Cameroons on the African coast, as suggested, although very inadequately by the maps shewn overleaf. We cannot dismiss this close fit as a pure coincidence, for not only are the coastlines similar on the two sides of the Atlantic, but also the mountains, the rocks, and even the fossils. For these reasons geologists have for long suspected that the two continents had once formed a single mass; the new theory provides an explanation of how they became separated. If North America is now towed still farther to the east, it will fit quite well on to Europe, New England fitting on to our Old England. Wegener believes that all the land which stood out above the sea some hundreds of millions of years ago can be fitted together to make one continuous continent, which would then cover about a third of the face of the globe.

Apart from all theories or hypotheses, we know that the

heights of mountains, and even of continents, are not permanent fixed quantities. When we climb a mountain, we expect to see an occasional stone rolling down, but we should be surprised to see one rolling up. Rain, snow, ice and even wind are continually splitting and loosening the rocks high up on the mountain side,

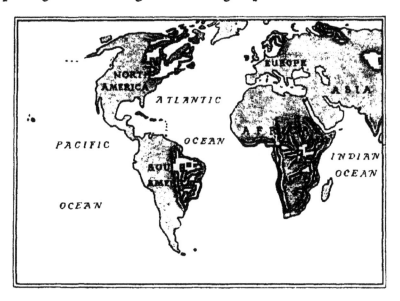

Fig. 7a. The continents in their present positions, with the vast rocky masses that are believed to have existed on them in primitive times.

until finally great pieces break loose and roll down to the bottom, so that huge boulders, piles of scree, and a general rocky detritus form familiar features at the foot of a mountain. Fig. 8 (facing p. 28) shews the accumulation of scree at the feet of the quite small mountains which lie on the east side of Wastwater in Cumberland. On the tops of higher mountains snow falls and turns into glaciers, which slowly flow down into the valleys and bring enormous quantities of large stones and powdered rock with them. On the foothills rain falls, and makes the mountain torrents dirty with particles of earth which have been washed off

Fig 7b. Wegener's theory supposes that the primitive continents did not occupy their present positions, but were packed together to form continuous land.

the hillside, and are now being carried down to the sea. Everywhere we see the substance of the mountains being transferred to the bottom of the sea, a process which tends to lower the heights of the mountains and raise the level of the sea-bottom.

The theory of isostasy suggests that this general levelling may be compensated, at least in part. For as the substance of the mountain is washed away, the mountain loses weight, and so floats higher, regaining part of the height it had lost. The sea-bottom, on the other hand, oppressed by the extra weight of the silt and sediment which the rivers have carried down to it, sinks deeper and in this way loses part of the additional height which the deposition of silt and sediment would otherwise have given it.

These incessant readjustments of level, and other causes as well, result in all sorts of settlements and upheavals in the surface layers of the earth; a whole continent may be thrust below sea-level, or a new continent may arise from the ocean-bottom and form dry land. As far back as the sixth century before Christ, the Greek Xenophanes recorded that sea-shells had been found far inland, and even high up in mountains, and that imprints—i.e. fossils—of fishes and seaweed had been found in the quarries of Syracuse. We need not travel as far either in space or in time as ancient Greece to find evidence of this process; it surrounds us everywhere, and especially in the chalk hills around London. These are full of fossils and the shells of tiny marine animals, which shew that they once formed the bottom of a fairly deep sea, rather like the middle of the present Atlantic Ocean. We also find submerged forests and even remains of life under the sea round our British coast.

The removal of rock and soil from the mountain tops is described as "denudation", and its deposition in valley bottoms and river beds as "sedimentation". This deposition formed the sedimentary layers that we have already compared to layers of dust outside the main body of an apple. If it were not for repeated settlements, upheavals, and general readjustments of level, the different layers of sediment would be deposited perfectly evenly, and would lie horizontally one over the other like the pages of a book on a level desk. Indeed there are large areas of the globe where the different layers lie perfectly regularly over one another in precisely this way—a large part of Eastern Canada, a large part of Eastern Siberia, a large part of the Baltic coast and Western Russia, and the relics of the ancient continent known as Gondwana Land, which included most of eastern South America and South Africa (see fig. 7 b, p. 27), Arabia and India. On a far smaller scale, we may often see the different layers of rock or sand lying one above the other in perfectly level layers in a railway or road cutting, or in cliffs on the sea-shore, or on an inland mountain, such as that shewn in fig. 9 on Plate III. The geologist describes these layers as "striations". Fig. 10 on Plate IV shews an immense cutting of this kind—the Colorado Canyon in North America. This cutting is not of the kind which man makes in a few days, but which Nature makes in millions of years. It was made by the Colorado River slowly eating its way down into the earth year after year and century after century, washing away the soft earth as it did so and carrying it down to the sea. We see layer after layer exposed, to a depth of more than 5000 feet; most of the layers look fairly horizontal to the casual glance, but the trained geologist finds evidence in places of elevation, of tilting, and even of subsidence under the sea.

In other places there may be no general subsidence or elevation, but the earth's crust may crack locally, and one side of the fracture may slide past the other, so that the striations no longer run on continuously but form what is known as a "fault" (fig. 13, facing p. 36).

If a river like the Colorado River had cut its way through England, we could have replaced fig. 10 by a view of the different layers of rock under our English soil. We have no river

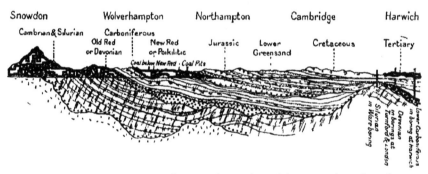

Fig. 11. Diagrammatic section shewing the geological layers under a line drawn across Great Britain from Snowdon to Harwich, a distance of about 200 miles We see how the underground layers have been folded, and uplifted parts removed.

to provide us with such a view, but a study of the surface formations, supplemented by such knowledge as can be obtained by borings and diggings here and there, enables the geologist to construct a diagrammatic map of the soil under England, which is almost as good and reliable as a real section, such as might be exposed by actual cutting. Fig. 11 shews a map of this kind which exhibits how the strata lie under a line drawn across England and Wales from Snowdon to Harwich;—a line running approximately west to east for over 200 miles.

However things may be in other parts of the world, we see that in our own country the geological strata no longer lie flat like the leaves of a book, but have been tilted and crumpled by rearrangements of level of the kind already described. Clearly there has been a general tilting movement which has depressed the east and raised the west, although marked local variations occur in places. Near the eastern end of our line, for instance, a subterranean upheaval has brought to within a few feet of the surface rocks which would otherwise have been many thousands

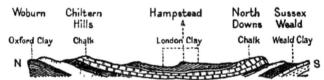

Fig. 12. A section similar to that shewn in fig. 11, but running under London, approximately north to south, from Woburn to the Weald of Sussex.

of feet down. This tilting of the strata has not brought about any corresponding slope in the surface of the land, since those parts which would have stood highest have all been washed away. If we continue comparing the different strata to the leaves of a book, we must not only imagine that the whole book has been very badly twisted and battered, but also that large parts of it have been rubbed away. There has been so much denudation that the surface of the land forms an almost horizontal section of the tilted book, and by merely crossing England from east to west, we obtain samples of the various pages in turn, all in their proper order.

In fig. 11 the successive geological layers had to be shewn

on so small a scale that many details were omitted. Fig. 12 shews a more detailed map of a shorter section, about 70 miles in length, passing under London from north to south. In this we do not notice any very pronounced tilting either to the north or to the south; the most conspicuous feature now is a general bending of the leaves of our book. We see that London is lying on clay, below which is a crumpled layer of chalk with a uniform thickness of about 650 feet. We have already seen how this once formed the bottom of a deep sea, and it is easy to reconstruct the story of subsequent events. First a subterranean upheaval so that the level sea-bottom becomes undulating dry land; then water streaming through this chalky land from still higher clay land to the west; then a broad river gradually depositing clay sediment which partially fills up the river bed but leaves the higher chalky hills unaffected; next a settlement of primitive men on some convenient land near the banks of the river; finally the London we know, lying on clay but surrounded by chalk hills which extend on the south from the white cliffs of Dover to the hills of Guildford and the Hog's Back beyond, until they emerge north of the Thames at Henley, and extend through the Chiltern Hills to the chalk downs of Hertfordshire and Cambridgeshire.

Even in this map the distortion of geological layers is still on a fairly extensive scale, but precisely similar distortions can often be seen in samples of rock only a few inches across, and sometimes even in the minutely thin sections of microscopic slides.

Now the real interest of this book of tilted pages is that it is in effect a history book. No matter how tilted and crumpled its pages may now be, each was originally deposited throughout its whole length and breadth at one definite date, or at least within the limits of one definite epoch, and contains buried within itself a history of that epoch.

To understand this, let us turn our thoughts from the Thames to the Nile. Every year the Nile floods the fields of Lower Egypt, and when it recedes, the level of Egypt has been raised a fraction of an inch by the sediment that the floods have left behind them. If we dig down a foot into this sediment, we shall find objects that had been lost or abandoned on the soil of Egypt 500 years ago; 4 feet down we come upon objects that must have lain there since the birth of Christ. The soil of Egypt forms, in effect, a stratified record of the history of Egypt; to turn over the pages of history, we need only dig down into the soil. Coins and inscriptions tell us the features and achievements of the kings; common implements, weapons, and tools, bring before us the lives of the people.

Other parts of the earth's surface can be treated in the same way, except, of course, that a foot of sediment will not always represent 500 years of history: sediment is deposited in different places at very different rates. Also, the various pages of history no longer follow one another in regular succession as we dig down; in many places the pages of our book have been crumpled and generally disarranged by faults, settlements, and upheavals of the earth's crust. This is really very fortunate, since if the pages had all lain flat on top of one another, we should have had

to dig more than 100 miles down to reach the lowest page. With
things as they are, we may often reach the lower pages by boring
only a short distance down, and sometimes merely by walking
over the earth's surface.

As we review these pages in turn, even if only by walking
over the surface of the earth with a discerning eye, we are again
in effect turning over the pages of history—no longer of Egypt,
but of the earth itself—and passing through the records of its
various civilisations. First we read of highly civilised men who
left behind them coins and inscriptions; then of earlier men who
left implements and weapons of metal and flints, sometimes with
the bones of the animals they hunted. Lower down still, we
come to the records of ape-like men who had nothing to leave
but their own dead bodies, which have since turned into skeletons. Then we come to the world before man was, a world in
which we find only the remains of weird animals and ungainly
monsters; then only fossils of reptiles, fishes, and plants. Finally
a world of lifeless soil, water, and rocks.

It would be thrilling enough to know the sequence of events
in the remote past, even if we could not say precisely when they
occurred—many of us do not find dates the most exciting part of
history. Yet perhaps the reason of their unpopularity at school
was that we were expected to remember them with such mathematical precision. It was quite interesting to read how, some
seven centuries ago, King John was made to sign Magna Carta
when he did not want to, but rather a bore to have to burden
our memories with the number 1215.

Now the physicists have recently found a means of dating the pages of the earth's history book, and this in the nicest possible way—with dates that are just exact enough to be interesting, but not exact enough to be tiresome.

We have all seen watches whose hands are visible in the dark because they are painted with radium paint. The hands appear to glow with a steady light, but careful examination with delicate instruments shews that the light is not really as steady as it looks; it results from myriads of separate explosions, each of which is caused by the death of a single atom of radium—perhaps we ought rather to say by the transformation, for the atom of radium does not entirely cease to exist, but leaves an atom of a special kind of lead behind it as a record of its former existence. Now this transformation of radium into lead invariably goes on at an absolutely uniform rate, which can of course be measured in the laboratory. Thus, if we could measure how much radium and how much lead there was in the hands of one of these watches, we could tell how old the watch was.

We can determine the ages of the rocks of the earth's crust in a similar way.

When thin sections of such substances as mica and tourmaline are examined under the microscope, they sometimes shew what is described as a "pleochroic halo"—a display of concentric rings such as is shewn in fig. 14 on Plate V (p. 36). At the centre of the halo there is always a minute speck of radioactive substance which decays in the same uniform unalterable way as radium,

3-2

only much more slowly; the substance may be uranium or thorium, or perhaps a mixture of both. The rings of the halo have been caused by the disintegration of this radioactive substance. Similar rings can be produced artificially in the laboratory, so that the mode of their formation is well understood. It is found that the colour of the halos deepens with advancing age, and it is often possible to estimate the ages of rocks simply from the general appearance of the halos they contain.

There are, however, a great many kinds of rock which contain uranium or thorium without shewing any halos. In such cases chemical analysis will tell us exactly to what extent these substances have disintegrated, and knowing this we can estimate the ages of the rocks—just as we might do with the hands of the watch. For instance, a large number of samples of the pegmatite rocks of Eastern Canada have been analysed, and all agree in shewing that the rocks solidified about 1230 million years ago. Other rocks shew ages which are even greater, but they are never much greater, and usually cannot be fixed with anything like the same degree of precision.

Thus we may say that the pegmatite rocks are the earliest page of our history book on which a definite and unmistakable date is written. On this page we read that 1230 million years ago the earth had a solid crust over which rivers flowed, washing mud and sand down to the sea. Pages even lower than this, undated, tell of earlier processes of cooling and solidification. We cannot say how long these earlier processes lasted, but they must have lasted through many millions of years, so that it hardly

PLATE V

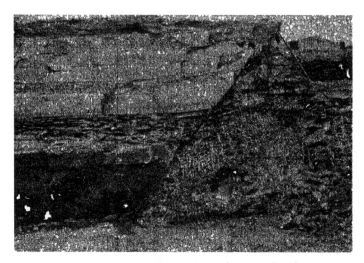

Fig. 13   Striations in sandstone at the Clydesdale iron and steel works, Mossend, Lanark, broken by a "fault".

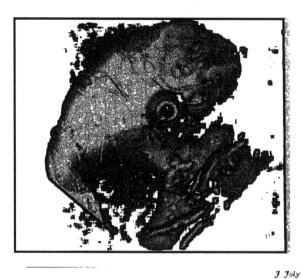

J. Joly

Fig 14. The halo produced by a minute speck of radioactive substance in mica. A microphotograph with more than 200 magnification

PLATE VI

Harvard University Press

Fig 15 The oldest known fossils—microscopic algae The photograph shews a slice of rock cut thin enough to be transparent and magnified about 190 times.

" Universal History of the World"

Fig. 16 Fossils found in Cambrian rocks These represent very primitive types of life—sponges, jelly-fish, star-fish, earth-worms and coral—some of which still exist in not greatly altered forms

seems possible that the earth should be less than 1500 million years old.

The earth cannot be enormously older than this, since, if it were, its radioactive substances would all have disintegrated away by now, and the phenomenon of radioactivity would have been entirely unknown to us, as it probably will be to beings—if any such there be—who will inhabit the future earth of a million million years hence. Unless radioactive substances have a capacity for renewing themselves in some way which is at present unknown to us, a detailed discussion shews that the age of the earth cannot possibly be more than about 3400 million years and is probably considerably less.

Somewhere between these limits—1500 million and 3400 million years—the age of the earth must lie. It is safest to confine ourselves to round numbers, in which case we may think of it as about 2000 million years—more than a hundred thousand times the whole length of recorded human history, and more than a million times the length of the Christian Era. It is not easy to realise what such figures mean. We may perhaps best visualise a million as the number of letters in a fair-sized book—a book, say, of 500 pages, with 330 words on each page, and an average of six letters to a word. If we take such a book to represent the age of the earth, then the whole of recorded human history will be represented by the last word in the book, and the whole Christian Era by less than the last letter. Within the space of this last letter the Roman Empire rose and fell; Christianity has spread over the face of the earth; the countries of Western Europe have changed from the savage countries described by Caesar to what they are to-day; more than sixty generations of men have lived and died. The whole of your life or mine will be represented by less than the final full stop, or the dot on the smallest "i" in the book.

If we want to read further back in time than this last word, our history book must be the earth's crust, with its layers of rock and soil for pages. Many of these have become crumpled in the course of time, but they are still arranged in the right order, and here and there a few of them are dated. Let us imagine that we straighten them out, and then turn them over and read what we can of the history of our earth.

We begin something like 2000 million years ago, and for hundreds of millions of years we watch a lifeless earth cooling and settling down. Page after page tells us only of geological activities, until, somewhere in the region of the "1230 million years ago" page, we begin to read of sediment containing traces of carbon. Some geologists consider that this provides presumptive evidence that some sort of life, possibly of the humblest and lowest description, existed in the sea; life, then, had already arrived on earth. Again we turn over page after page, and read only of geological events—steady sedimentation, varied only by cataclysms and upheavals—until at last, somewhere between 1000 and 500 million years ago, we come upon fossil remains, mere specks embedded in the rocks (fig. 15) which the geologist interprets as definite remains of life, although life of the most primitive kind. Again, long aeons pass before us until,

PLATE VII

Universal History of the World"

Fig 17 Fossils of the Silurian period, about 450 million years ago. This is sometimes described as the Age of Sea Lilies, because sea plants such as those shewn grew at the bottom of the sea in such dense masses that their fossil remains now form thick beds of limestone The "sea lilies" are not true plants, but are more like our star-fish or sea-anemones

PLATE VIII

W. F. Swinton, "Monsters of Primaeval Days"

Fig. 18 *Dimetrodon Gigas*, a huge and clumsy carnivorous lizard, 9 feet in length, which is believed to have inhabited North America about 250 million years ago

Universal History of the World

Fig. 19 *Cacops Aspidopharus*, one of the reptiles which made its home on dry land in the great drought of about 200 million years ago

about 500 to 400 million years ago, life becomes both more complex and more abundant. We even find fossils of worms, jelly-fish, and other rudimentary forms of life, which were not enormously different from those existing to-day (fig. 16).

Again, years roll on in their millions, until we open a page of our book on which the pictures—the fossils—look very like the plants of to-day. They look like plants but are not, for they lived at the bottom of the sea; they were more like our sea-anemones, or even star-fish, than plants (fig. 17). Yet shortly after this time life begins slowly to invade the land, and we come upon the first fossils of true grasses and fern-like growths. As the land vegetation multiplies we see the earth gradually assuming something like its present appearance. The roots of the grasses fix the particles of sand and earth to form a solid soil, while animals appear to feed on the vegetation, and others, in due course, to feed on them. This was the beginning of the era when huge reptiles dominated the earth. Typical of the earlier of these was the *Dimetrodon Gigas* (fig. 18), a huge carnivorous lizard, which is believed to have lived in North America about 250 million years ago.

The humbler forms of life, such as the worms, jelly-fish and sponges shewn in fig. 16, have survived, without any very great alteration, from that far-off period until to-day, but the more complicated forms of life were destined to undergo many changes.

For, as we read on in our book, we come to pages on which the geologists have written "Permian Era" and "Triassic Era", and the physicists "about 200 million years ago". On these we read of great mountain upheavals which completely altered the face of the globe. In the northern hemisphere most seas, including the present Atlantic and Indian Oceans, became dry land, and only part of the present Pacific Ocean remained as an ocean. In the southern hemisphere, the great continent which geologists describe as Gondwana Land emerged from the sea, to occupy the whole stretch from eastern South America through Africa to Australia. The geologists shew us small depressions in the rocks, in which the fossils of fish are packed like sardines in a tin, as though they had spent the last moments of their life crowding together where they could take advantage of the last few drops of water before these evaporated. With so little sea to give moisture to the air, the rainfall naturally decreased, and the greater part of the world became a desert. In particular we read of the seas of Northern Europe contracting into salt lakes, which became more and more salty as the drought increased in intensity, until they finally dried up altogether, leaving deposits of solid salt such as we now find in Cheshire and Staffordshire.

Then the drought begins to pass, but many forms of life fail to reappear in the later pages of the great book. They must have perished in the drought, and indeed it is obvious that only those which were able to adapt themselves rapidly to new conditions could hope to survive. An example is shewn in fig. 19 (p. 39) (*Cacops Aspidophorus*, or the Grim-faced Shield-bearer), an enterprising but not beautiful reptile, who somehow managed to find a living on dry land after the seas had dried up.

Next we come to pages marked "Jurassic Era" bearing dates of from 150 to 100 million years ago. These tell us of the seas again flowing over the parched deserts, of moisture returning, and of the earth again becoming hospitable to life. Such reptiles as have survived the drought now distribute themselves again over sea and land, and even invade the air, for we are now coming to those pages of our history book in which the fossils of winged creatures first appear—weird, ungainly birds, some with teeth, some with toothless beaks.

Many of the animals that inhabited the earth at this time were failures and misfits, unsuited to survive in the great struggle for existence, although a number of them contrived to live for a great many years before this fact was relentlessly borne in upon them. Figs. 20, 21, 22 and 23 shew examples of four such creatures, who lived in North America from 80 to 100 million years ago, and have since become extinct.

Fig. 20 portrays the *Triceratops*, who is typical of a whole class of animals who trusted to defensive armour. He had three horns, each many feet long, and when he was attacked, he only needed to stand with his back to the wall and wait until the foe had impaled itself on his horns. He was a huge creature, about 25 feet in length, and standing 9 feet high. He was still a reptile of a sort, and his female laid eggs of vast size.

Fig. 21 shews another creature in much the same class—the *Scolosaurus*, or Thorn-reptile, a member of a family which has been described as consisting of "the most ponderous animated citadels the world has ever seen". His method when attacked was probably to flatten himself on the ground and wag—or perhaps swish—his tail, which, as you see, ended in an immense knob of bone, rather like the spiky maces which the Crusaders used to wield. In those days, the tactics of defence and attack seem to have been equally rudimentary, and did, not call for a high level of intelligence; *Triceratops* had a skull 6 feet long, but his brain was the size of a kitten's.

Fig. 22 shews a Pterodactyl—*Pteranodon occidentalis*—a huge bird-like reptile with a wing-spread of about 18 feet. He was one of those unhappy creatures who can do quite a lot of things in a rather futile sort of way, but nothing really well. He had wings which were probably rather too weak to raise his heavy body through the air, so that he could not fly well, and legs which were too weak to carry his great weight on land, so that he could not walk well. Possibly he could not run at all. He could not even sit down very well, since his elbows must always have got in the way, unless he sat perched on the top of a rocky peak. Scientists picture him as spending dreary days trudging laboriously to the top of a hill or cliff, then launching himself into the air currents, floating through these like a glider until he could swoop down upon his prey, and then starting to climb the hill again. I think we may properly feel sorry for him—his life must have been like an endless repetition of learning how to ski without a funicular.

Fig. 23 shews the *Diplodocus*, one of the largest animals that ever lived on earth. He was about 30 feet high, and about 90 feet long, so that a single *Diplodocus* must have weighed as

W I Stanton   The Dinosaurs

Fig. 20. *Triceratops Prorsus* lived in North America about 90 million years ago. It lived on vegetation, was 25 feet long, had a skull as large as an elephant's, but a brain inside it no bigger than a kitten's.

"*Universal History of the World*"

Fig. 21. *Scolosaurus* lived in Canada about 90 million years ago. Its only weapon, either for defence or attack, appears to have been its knobby tail.

Fig. 22. *Pteranodon occidentalis*, a winged reptile, which inhabited North America about 90 million years ago. Its wings had a spread of about 18 feet, and consisted of thin webs of skin stretched from the fifth fingers of the arm to the hind legs

Fig. 23. *Diplodocus*. This immense reptile lived in North America about 90 million years ago. Most of its great length of 90 feet was taken up by its long neck and very long whip-like tail.

much as a whole family of elephants—father, mother, children, and perhaps several uncles and aunts as well. A good specimen would perhaps have turned the scale at anything from 40 to 50 tons. He was so heavy that his legs could hardly support him on land, and so preferred to live in marshes, where his long neck came in useful at feeding time; indeed, he really needed the buoyancy of water to ease his weight if he was to live in any sort of comfort. Fig. 24 shews another huge, and even more un-gainly, reptile—the *Cetiosaurus*, or Whale-reptile. He lived on our own side of the Atlantic, and many skeletons have been, and still are, found in English stone-quarries. His length was about 60 feet, and like his American relation, the *Diplodocus*, he found it comfortable to ease the strain on his legs by living in the water. Let us not ridicule the disabilities of these un-happy creatures, for when we reach Jupiter and Saturn we shall find ourselves in an exactly similar predicament, and may have to take similar precautions if we are not to collapse under our own weights.

Just as we shall find that we are unsuited for anything more than a very brief visit to Jupiter or Saturn, so these misfits were unsuited to hold their own for long in the struggle for existence on earth, and were forced in time to yield to their more agile competitors—the smaller mammals, and finally man—who trusted to activity and intelligence rather than to heavy armour or brute size and weight. The big and heavily armoured reptiles gave way for precisely the same reason for which the heavily armoured soldier of the Middle Ages has given way to the unarmoured soldier of to-day, the reason for which fortresses and battleships are yielding place to tanks and torpedo-boats, and the airship to the less unwieldy aeroplane.

After these creatures had become extinct, we come to the age of mammals and of creatures who were in general more like those of to-day. Fig. 25 shews the *Arsinoitherium*, who lived in Egypt some 25 million years ago. He was far smaller than the monsters of the preceding age, and yet was as big as an average rhinoceros, or small elephant. He makes me think of Mr Kipling's "Just-So" story of how the elephant got his trunk. The baby elephant, you remember, was far too inquisitive, or so his family thought, about the facts of natural history; they particularly disapproved of his continually inquiring what the crocodile had for dinner. Finally he asked a crocodile basking in a swamp, who told him to lean down, so that he could whisper the answer in his ear. When he did this, the crocodile very un-kindly and treacherously caught on to his nose, saying "Baby elephant to-day", and pulled and pulled until the nose was elongated into the trunk which every elephant now possesses. The *Arsinoitherium* looks rather like baby elephant must have looked half-way through the process, although the protruding part of his face is not really a trunk or nose, but consists of two pointed horns of bone which grew just above his nose. He had also two similar but smaller protrusions over each eye, and must have looked terrifying and fantastic to the last degree.

Fig. 26 shews us a smaller, but much more fearsome, creature —the *Machaerodus*, or Sabre-toothed Tiger, who inhabited Asia

PLATE XI

British Museum (Natural History)

Fig 24 *Cetiosaurus*, the Whale-reptile This is a British relative of the *Diplodocus* shown in fig 23 It was about 60 feet in length, and correspondingly heavy

W E Swinton, "Monsters of Primaeval Days"

Fig 25 *Arsinoitherium* inhabited Egypt about 25 million years ago. In appearance it is somewhat like an elephant, but in many other respects is more like a rhinoceros.

PLATE XII

"Universal History of the World"

Fig 26 *Machaerodus*, the Sabre-toothed Tiger, was a distant relative of the ordinary cat, but was not a tiger at all.

W. E. Swinton, "Monsters of Primaeval Days"

Fig 27 *Megatherium*, the Giant Ground-sloth, was a vegetable eater, about 20 feet long, and 12 feet high when it sat up on its hind legs to reach the branches of trees

and Europe from 1 to 10 million years ago. He was about the size of a large tiger or lion, and his mouth contained two really terrific teeth, immensely long and thin—sharp in front and saw-like behind—which look very formidable, and yet seem to have formed an effective obstacle to his either closing his mouth or eating his food; indeed no one seems quite to understand why he did not die of starvation.

Fig. 27 shews us the *Megatherium*, or Giant Ground-sloth, who lived in South America within the last million years. You can judge of his size from the man who is with him in the picture. This huge creature was quite harmless; he was perhaps hunted by man, and possibly even kept as a domestic animal, for the remains of one were found in the same cave as those of a man.

These huge sloths have been extinct for a long time, but the man is our own ancestor. Somewhere within the last million years ape-like mammals developed—or perhaps suddenly changed—into man, and we are the result. In comparison with a single lifetime, a million years seems almost interminable; in comparison with the total age of the earth, it is merely a small fraction of a fraction. In fig. 28 (p. 46) the principal epochs of the earth's history are drawn to scale. The million years or so of man's abode on the earth are represented by less than half the thickness of the thin line at the top of the diagram.

Yet even through this tiny fraction, man was mainly un-civilised, living little better than the beasts he hunted. We glance over hundreds of thousands of years of human history, and see only savages living in caves like animals, fighting with

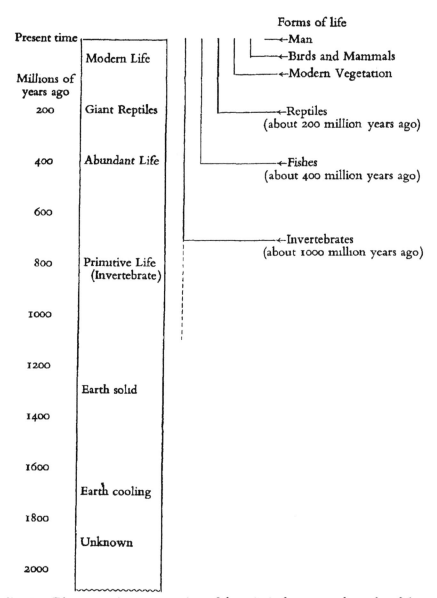

Fig. 28. Diagrammatic representation of the principal events and epochs of the past history of the earth. (The events and dates are highly conjectural.)

animals, and perhaps crying like animals. Then, perhaps 100,000 years ago, he acquires a new capacity for speech; he becomes able not only to plan and devise, but also to exchange his ideas with his fellow-men, and communicate his plans to them. This gives him an almost unchallengeable ascendancy over all other animals, and henceforth his progress is rapid. Perceptible change is no longer a matter of millions of years, thousands suffice, then centuries—now, almost single years. Human life has changed more in the last 50 years than reptile life did in 50 million years in the Jurassic and Permian Eras.

CHAPTER II

THE AIR

Let us leave the earth, in which we have burrowed for long enough, and turn our thoughts, and our eyes, upwards.

We all know what we may expect to see—the sun, the blue sky, and possibly some clouds, by day; stars, with perhaps the moon and one or more planets, by night. We see these objects by light which has travelled to us through the earth's atmosphere, and if we see them clearly, it is because the atmosphere is transparent—it presents no barrier to the passage of rays of light.

Perhaps we are so accustomed to this fact that we merely take it for granted. Or perhaps we think of the atmosphere as something too flimsy and ethereal ever to stop the passage of rays of light. Yet we know exactly how much atmosphere there is, for the ordinary domestic barometer is weighing it for us all the time. When the barometer needle points to 30, there is as much substance in the atmosphere over our heads as there is in a layer of mercury 30 inches thick. This again is the same amount as there would be in a layer of lead about 36 inches thick, for mercury is heavier than an equal volume of lead in the ratio of about six to five. To visualise the weight of the atmosphere above us, we may think of ourselves as covered up with 144 blankets of lead, each a quarter of an inch in thickness. We should hardly expect to see through 144 blankets of lead, so that it begins to look rather surprising, and perhaps something of a lucky accident, that we can see through the equally substantial atmosphere.

If so, it is a piece of luck which many of the planets do not share with us. When we examine the other planets from the earth, we find that most of them are covered in with opaque atmospheres which prevent our seeing their surfaces at all. Thus we may be forewarned that when our travels take us to these planets, we shall not be able to see through their atmospheres to the sky and stars above.

Let us, however, look into this question of opaqueness and transparency to light in some detail. We know that light, like all other forms of radiation, consists of waves, and we know that waves may be either long or short. In the case of sea waves, for example, there are the long breakers, perhaps hundreds of yards long, which rock even the biggest of ships; there are also little ripples only a few inches long, which have no appreciable effect on big ships, but rock row-boats—or perhaps they are too small to affect row-boats, and only disturb still smaller objects, such as pieces of cork or seaweed. It is the same with waves of light; some are long and some are short, and waves of different lengths affect objects in different ways.

Now the radiation which the sun emits contains waves of almost every length all mixed up together, although some lengths of wave occur only in minute quantities. Our eyes are not sensitive to those kinds of waves which the sun sends out only in small amounts, and neither are they sensitive to certain other kinds of waves which are emitted in abundance by the sun, but fail to reach our eyes because our atmosphere refuses them passage. If such waves should suddenly begin to penetrate our atmosphere in abundance they would burn us, and we should turn first brown, then black, and would shortly die, but our eyes would never see the light that was killing us. As a general rule, our eyes are sensitive only to light of those wave-lengths which reach us in abundance—that is, in brief, to waves of the kinds which make up ordinary daylight.

Perhaps this is hardly surprising. We are the offspring of millions of generations of ancestors, whose organs, including their eyes, have been slowly and gradually adapting themselves to their environment for hundreds and millions of years. For this reason, we seldom find either animals or men encumbered with organs that serve no useful purpose. When an organ is no longer needed, so that it falls into disuse and becomes merely a useless burden, it gradually disappears; or if it does not, the animal burdened with it may disappear, as did the heavily armoured reptiles whose acquaintance we made in Chapter 1. Eyes that were sensitive to light such as never reached them from the sun would have been mere encumbrances, both to animals and men, and if the human race had ever had such eyes, they would certainly have disappeared by now.

As our bodies have gradually developed through millions of years, our lungs and blood have adapted themselves to the quality and quantity of the earth's atmosphere and our skins to its climate—black skins for the tropics, white for the temperate zone, and so on. So our eyes have in all probability adapted themselves to daylight, and it is not mere luck that they are mainly sensitive to just those kinds of radiation which reach them in abundance. When we arrive on Jupiter, we shall find that we cannot see through its clouds. If, however, we had lived on Jupiter for thousands of generations, our eyes might have adjusted themselves to some special kinds of waves which pass through the clouds of Jupiter. We might have been saying how fortunate we were to live on Jupiter, with its beautifully transparent atmosphere, and pitying the inhabitants of other planets, such as the earth, who were shut in by opaque clouds.

As our knowledge of everything outside the earth comes to us in the form of radiation, and especially of light, it is very important for us to know the properties of different kinds of light and radiation in some detail. When we look at a rainbow, or a patch of dewy grass in sunlight, we see a profusion of colours. We know that if the sun went out, or even disappeared temporarily behind a cloud, we should no longer see the rainbow or coloured dew. This proves that the light we see started in the first instance from the sun. But it has not come to us by the most direct path—it reaches us from the wrong direction for that. It has been reflected into our eyes by little globules of water—the raindrops of the shower, or the dewdrops on the lawn—and in passing in and out of these tiny drops of water, it has been broken up into the assortment of colours that we see. There are of course more effective ways of breaking up the sun's light; we can break it up by letting it pass through a glass prism, or even a bottle of

4-2

water, or—most effectively of all—by using the very sensitive instrument known as the spectroscope.

When the sun's light has been broken up by any of these ways, it appears as a band of varied colours, with red at one extremity and violet at the other. This band is called a "spectrum". There are other colours in between, so that the complete succession of colours runs: red, orange, yellow, green, blue, indigo, violet. If we break up any other kind of light, we shall get another spectrum, but whatever kind of light we use the colours invariably follow one another in the order just mentioned. The reason is that the different colours of light are produced by waves of different lengths, and that in every spectrum the different colours of light are arranged in order of their wave-lengths.

We shall obtain a simple proof of this if we break up the light by another instrument—the diffraction grating. This is merely a metal plate on which thousands of parallel lines have been scratched at exactly equal distances from one another with a diamond or other hard point. When light falls on the surface of the metal, this series of scratches picks out the waves of different lengths and reflects them in different directions, thus sorting them out according to their length much as a potato sieve sorts out potatoes by their sizes. The distance between the successive scratches corresponds to the mesh of the sieve, and, when we know this, we can calculate the lengths of any particular waves from the directions in which they are reflected. When light is broken up in this way, the different wave-lengths again form a spectrum, in which the colours appear in precisely the same order as before. But we need no longer grope in the dark as to the meaning of this order; we now see at once that the different colours of light are produced by waves of different lengths and that in the spectrum the colours are arranged in order of wave-length. Actual measurements shew that red light has the longest wave-length, about 33,000 waves going to an inch. As we pass through the other colours—orange, yellow, green, etc.—in order, the wave-length continually decreases, until we come to violet light with 66,000 waves to the inch.

Sound also consists of waves, although of a very different kind; they need air to travel through, and are about a million times as long as light waves. Just as different colours of light are produced by light waves of different lengths, so sounds of different pitch are produced by sound waves of different lengths. For instance, middle C on the piano has a wave-length of 4 feet, while treble C has a wave-length of 2 feet. When one sound has just half the wave-length of another, we say it is an octave higher in pitch. In the same way when one colour of light has just half the wave-length of another, we may say by analogy that it is an octave higher in pitch. Thus, as violet light has just half the wave-length of red light, we may say that violet light is an octave higher in pitch than red light. Indeed, we shall not go far wrong if we think of the seven colours of the spectrum as the seven notes of a scale, red being C, orange D, yellow E, green F, and so on. We have already seen that all the visible spectrum lies within one octave. Our ears can hear eleven octaves of sound, but our eyes can only see one octave of light.

We have also noticed that the sun's radiation consists of far more than the one octave of light our eyes can see. Beyond the deepest violet light that we can see, there is a great deal of light that we cannot see; it consists of waves even shorter than those of violet light, and is spoken of, in a general way, as "ultra-violet radiation", or even as "ultra-violet light". It does not affect our eyes for precisely the same reason for which tiny ripples on the sea do not affect a big ship—its waves are too short. But it does very powerfully affect photographic plates, and if the retinas of our eyes were made of substances similar to the emulsion of photographic plates, we should see ultra-violet radiation.

Out beyond the red end of the spectrum also, there is a great deal of radiation which our eyes cannot see; this consists of waves longer than those of red light, and is commonly described as infra-red radiation. When a solid object is heated up—as for instance a horseshoe at a blacksmith's forge—it glows at first with a dull red light, then, as it gets hotter and hotter, it becomes bright red, orange, and yellow in turn. The act of heating it up causes it to emit radiation, and the hotter it becomes, the shorter the waves it emits. We may say that, as an object gets hotter, its radiation moves along the spectrum in the direction of shorter and shorter waves. We do not begin to see the object by its own light until its radiation has passed into the visible part of the spectrum, but long before this stage is reached it is giving out radiation in the infra-red part of the spectrum. Our skins, but not our eyes, are sensitive to this radiation; if we hold our hand near a hot horseshoe, we shall feel its radiation long before we can see it with our eyes. This shews that the infra-red radiation is of the nature of heat rather than of visible light. Ordinary photographic plates are not affected either by infra-red or by red light; for this reason we can use red light in the dark-room without damaging our sensitised plates. If the retinas of our eyes were made of substances similar to the emulsions used on ordinary photographic plates, we should not see red light at all, and should hardly be able to see yellow or green—only blue, violet, and the ultra-violet which our present eyes cannot see.

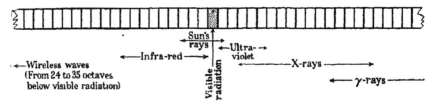

Fig. 29. The central part of the scale of radiation. Each section represents one complete octave of radiation, all being invisible except for the central octave which is shaded dark.

Although we can only see one octave of radiation with our eyes, scientists have found the means of studying as many as sixty-four octaves. Their scale of radiation is like a vast piano with sixty-four octaves, to all of which we are deaf except for the one octave of visible light (cf. fig. 29). Immediately above this one octave, going treblewards, we come to ultra-violet radiation. This makes its presence known by affecting photographic plates, and also by causing a number of chemical substances to "fluoresce", which means that when invisible ultra-violet light falls on them they emit visible light—as though they took the radiant energy and pushed it down several notes in the scale. Then, about ten octaves above the octave of visible light, we come to X-rays. Light substances are more transparent to these than heavy substances, so that when the rays are sent through a mixture of substances, the heavier substances cast deeper shadows than the lighter. Because of this property the surgeon can use these rays to photograph broken bones through the flesh, and they can also be used to examine an old painting even though a modern painting has been superimposed on the same canvas.

Above all these—very high indeed in the treble—come the γ-rays which are emitted by radium; and finally, thirty-two octaves above the octave of visible light, come certain of the constituents of cosmic rays, which can pass through many yards of lead.

In the other direction—down towards the bass—we come first to the infra-red radiation we have already described; the heat radiated from a hot flat-iron is about three octaves down, and that from a kettle of boiling water about four. Special photographic plates are now made which are sensitive to infra-red radiation, so that it is possible to photograph objects by this radiation in what appears to our eyes to be complete darkness. For instance, in fig. 35 (facing p. 76), we see a hot flat-iron photographed in a dark room by its own infra-red radiation.

Far below these—about thirty octaves below visible light—we come to waves which are more than a thousand million times as long as the waves of visible light. These are of special interest and importance, being nothing other than the waves used for radio transmission. Yellow light has a wave length of only about a forty-thousandth of an inch, but if we want to receive a wireless programme, we tune into such wave-lengths as 1500 metres, 342·1 metres, and so on. Except for everything being magnified a thousand-million fold, these waves have many of the properties of waves of light. For instance, the parallel wires of a beam station treat them almost exactly as the parallel scratches on the surface of a diffraction grating treat the waves of visible light. If we let light of one single colour fall on to a diffraction grating, we find that it is all reflected as a beam in one single direction, the exact direction depending on the wave-length of the light. In the same way, if radio waves of one single wave-length are sent through the aerials of a beam station, they all start off as a beam in one single direction—to India, China, Japan, or anywhere else we wish, according to the length of wave we use.

After this preliminary study of the properties of radiation and light, let us set out to study the atmosphere through which they travel. Perhaps most of us are disposed to think of it as a simple layer of transparent gas stretching up skywards, but scientific study shews it to be a structure of very great complexity. We shall get quite a good general idea of this structure if we think of it, like the earth itself, as a vast number of wrappings or layers, each enclosing the one inside it, until the innermost of all encloses a massive parcel—the solid earth.

The first layer of atmosphere enwrapping the earth is known as the "troposphere"—the sphere of change. Its thickness varies from about 5 to 10 miles at different times and places, being generally about 7 miles. Although this is only a small fraction of the whole thickness of the atmosphere, yet the troposphere contains nearly 90 per cent. of the total substance of the atmosphere. The explanation is of course that the atmosphere is far more dense in its lower layers, where there is a great deal of air above it, and so pressing it down, than higher up where it has but little weight to support. The troposphere is continually being agitated by winds and storms—hence its name—in contrast to the layer above, known as the "stratosphere", which is characterised by almost perfect calm; storms do not reach so high.

The atmosphere consists of a mixture of many kinds of gases, some light and some heavy. If it were left to itself for a sufficient time, the light gases would rise to the top, as the cream does in a basin of milk. Actually it is never left to itself for more than a few days at a time. We have already seen how the earth's rotation causes the trade-winds to blow, and what with this and all sorts of other winds and storms, the troposphere is more like milk in a butter churn than milk standing in a cream basin. This continual churning keeps the gases of the troposphere thoroughly mixed, so that its composition is the same throughout. As we know, it consists of four parts of nitrogen to one of oxygen, with other gases mixed in far smaller quantities.

The principal of these other gases is water vapour, and this has very special properties, since it alone of the constituents of the troposphere can condense into drops of liquid—raindrops— which then tend to fall earthwards in the form of rain or snow. We are all familiar with showers of water, but we have never seen showers of oxygen, nitrogen, or helium. Now water vapour is especially prone to form into these drops when the air is churned up by wind—hence the common belief that rain follows wind. We have said that churning up the air distributes the constituent gases uniformly throughout the atmosphere, but it now appears that we must make an exception in the case of water vapour— churning sends this down to the lowest level of all, the surface of the earth. After a time all the water which has fallen as a shower will evaporate, and in this way be restored to the atmosphere, but before it has ascended very high, another wind is sure to come along and shake it down again. Thus it is not surprising to find that the water vapour is not spread uniformly through the atmosphere, but is confined almost entirely to the lowest levels. Actually about one molecule in eighty is water vapour at sea-level, but at the top of the troposphere the proportion has fallen to 1 in 10,000. This means that practically all the water vapour of the atmosphere lives in the troposphere, which thus becomes the region of rain, snow and fogs. Ordinary rain clouds (fig. 31, p. 68) generally occur at heights which range from a few hundred feet to a mile or more, while the highest clouds of all, the fine-weather clouds known as cirrus and cirrostratus (fig. 30), are at an average height of 5 or 6 miles. Above the top of the troposphere there are no clouds of any kind.

The continual stirring up of the gases of the troposphere has one interesting and important result. When we apply pressure to a gas, it not only contracts in volume, but also rises in tempera-ture—compressing a gas heats it up, as we notice when we pump up a tyre. Conversely, releasing the pressure of a gas lowers its temperature. For this reason, the gas which issues forth from a cylinder of compressed gas always cools as it comes out; it may even freeze and come out in the form of snow—this is how many fire-extinguishers work. Now as air is carried upwards by the winds and storms of the troposphere, the pressure on it is released, so that it cools—just like gas coming out of a gas cylinder. If the same air is dragged down again by more winds and storms, its pressure is increased, and it gets hotter—just like the air in a tyre. For this reason, the upper layers of the troposphere are always colder than the lower. If we climb a mountain, or go up in an aeroplane, we find the air getting colder; if we go down into a mineshaft or valley bottom, we find it getting warmer.

If the atmosphere were a simple mass of churned-up air, we can calculate that its temperature would decrease by 29 degrees Fahrenheit for every mile of height. But many other factors must be taken into consideration, such as the heat of the earth, the sun's radiation, and the irregularities of the earth's surface. Actual observations with balloons shew that the temperature does in fact fall fairly uniformly with the height, but only at the rate of about 17 degrees Fahrenheit per mile. With a temperature of 60 degrees Fahrenheit at sea-level, the temperature 7 miles up will be about 60 degrees below zero. This is approaching the lowest temperature which has ever been recorded on the earth's surface, namely, 94 degrees below zero at Verkhoiansk, Siberia.

Early scientists imagined that anyone who went higher still would find the atmosphere getting colder and colder, until finally it became so tenuous that it could not properly be said to have a temperature at all. Then, in 1898, a series of balloons were sent up near Paris to determine the temperatures at great heights, and this view was found to be fallacious. The temperature was found to remain almost uniform after passing a height of from 7 to 10 miles, and sometimes even shewed a slight increase. The reason, as we now know, was that the balloon had passed out of the turmoil of the troposphere into the calm of the stratosphere; here there were no storms alternately to compress and rarefy the atmosphere and so to cool it up above and warm it down below. For a layer of gas which is repeatedly stirred up like the troposphere will develop a temperature gradient, but one which is left to itself, like the stratosphere, tends to assume a uniform temperature.

When we try to explore the heights of the stratosphere, we encounter much the same difficulties as in trying to explore the depths of the earth. The most obvious way of exploring the earth was to dig a hole, and either go down it ourselves, or send instruments down, to bring up a sample. But this only took us a very short distance down, after which we had to let waves do the exploring for us. In the same way, the most obvious method of exploring the stratosphere is either to go up ourselves in a balloon, or get a sample of air down in a balloon. Both these methods are in common use, but they do not take us very far. Up to the present, human beings have never gone higher than 13·7 miles, the height reached by a balloon sent up from Moscow in January 1934, and then they did not come down alive. The greatest height reached by a balloon without passengers was

23 miles attained by a balloon sent up from Padua. Heights
greater than these can at present only be explored by the passage
of waves. |Only one kind of waves, namely earthquake waves, are
available for the study of the earth, but there are three distinct
kinds available for the study of the stratosphere—waves of light,
waves of sound, and radio waves. Waves of all these three kinds
pass through the stratosphere, and can be made to bring a message
down with them, almost as a balloon filled with self-registering
instruments does.

The waves of light which pass through the stratosphere are of
course the radiation of the sun and stars. They bring with them
a message that they have been robbed of some of their constituent
wave-lengths in their passage through the atmosphere. Many of
the missing wave-lengths are in the ultra-violet part of the spectrum, and are found to be precisely those which cannot pass
through ozone. Thus it is natural to conclude that ozone is responsible for the robbery. Ozone is a specially heavy variety of
oxygen gas, having three atoms to the molecule in place of the
usual two. Popular imagination credits it with remarkable powers
in the matter of making our seaside resorts bracing, bringing the
glow of health to pallid faces, and so forth. Science knows nothing
of this, chemical analysis shewing that there is remarkably little
ozone either at our seaside resorts or anywhere else on land or sea.

It is found that the amount of ultra-violet radiation which
reaches the earth is not uniform, but varies with the position of the
sun in the sky. There is a quite definite relation between the two,
and this makes it possible to estimate the position of the ozone by

which the ultra-violet radiation has been absorbed. Recent investigations by Professor Dobson of Oxford and other scientists shew that most of the ozone lies within 25 miles of the ground, its average height being about 15 miles. The amount of ozone is extraordinarily small, its total weight being only that of a layer of paper about a two-thousandth of an inch in thickness—the thinnest of tissue paper. The sun's light can pass through miles of ordinary air without appreciable absorption, and yet this thin layer of ozone is enough to stop its ultra-violet rays from reaching us. In a sense, then, there may after all be some luck in our atmosphere being transparent to any light at all. For we might have been surrounded by an atmosphere whose different constituents shut off various parts of the sun's light as effectively as the ozone shuts off the ultra-violet rays, so that neither sunlight nor any other light could pass through it at all.

The ozone does not shut off all the ultra-violet radiation, and this is fortunate, since a certain amount of it is beneficial to us. It is said that miners, and other people whose work keeps them much below ground, find that their health is improved if they occasionally expose themselves for short intervals to artificially produced ultra-violet radiation. Children who have to all appearance been starving from want of adequate food, have sometimes been restored to health merely by letting this radiation fall on their skins, thereby producing the vitamin D which is essential to health. On the other hand, too much of the radiation may prove even more disastrous than none at all, and we sometimes hear of people dying from over-exposure to it.

The ozone layer controls the supply of ultra-violet radiation we receive from the sun, and, broadly speaking, gives us just about the amount we need. When we travel on other planets, we may find that their atmospheres let through too much or too little of this radiation to suit us and our health will suffer accordingly. Yet once again the reason that our own atmosphere appears to treat us so well is probably that our bodies have, after millions of generations, learnt how to get on with exactly what is meted out to them. If we had lived for millions of generations on some other planet, we might find the amount of ultra-violet radiation on earth intolerable.

Other lengths of waves are also missing in the radiation we receive from the sun and stars, particularly in the red and infrared parts of the spectrum. These omissions can, however, be traced to the presence of oxygen, water vapour and carbon dioxide, and so tell us nothing new about the composition of the atmosphere.

So much for what light waves have to tell us; we shall find that we can learn even more from radio waves. Unlike light waves, these do not come into the atmosphere from outside—except perhaps in quite inappreciable amounts—so that we must study the waves emitted by our own wireless stations. We have seen that these are of the same general nature as waves of light, except that they are thousands of millions of times longer. Being of similar nature, they have many properties in common with light waves. Both for instance travel in straight lines, and both are stopped by the solid body of the earth. Just as we can never hope to see round the earth, so it might have been anticipated that we should never be able to pick up a wireless signal emitted by a station at the other side of the earth.

For this reason, the earlier experimenters were greatly mystified when they found that they were picking up wireless stations at the opposite ends of the earth without difficulty; they can now pick up stations near their receiving sets by waves that have travelled twice round the world, and taken nearly half a second on their journey. Not only so, but everyone who has ever played with a wireless set knows that quite distant stations are often received better than near-by stations of equal power.

Gradually the conclusion was reached that radio waves were sent out in all directions, but that as soon as a beam reached a certain height above the surface of the earth, it was in some way bent back and returned to earth. If we found light waves behaving in a similar manner, we should conclude that somewhere up in the sky there was a gigantic mirror which reflected them back to earth. To some extent thick clouds behave just like such a mirror for light waves—for instance, when the sky is covered with clouds, the glare of the lights of London can be seen far out into the country. Yet the mirror which reflects radio waves back to earth must be something quite different from this—it must be completely transparent to ordinary light, because distant stations are often received perfectly well on a clear night.

It is known that an ordinary mirror reflects light because its surface is a conductor of electricity. The surface usually consists of quicksilver or metal, but air and other gases can also be made to

JTS

conduct electricity under special conditions, so that there is no reason why a mirror should not consist of air or gas. Generally speaking, gases conduct when they are "ionised", which means that electrons have been torn off from their molecules and so are free to move about and transport an electric current—which, incidentally, is exactly the process by which a film of quick-silver or a surface of metal conducts electricity. In 1902 two scientists, Kennelly in the United States and Heaviside in England, independently suggested that there must be a layer of ionised gas high up above the earth which acted as a mirror for radio waves, and turned them earthwards again. Since then, their conjecture has been amply confirmed, and the layer of ionised gas is known as the E, or Kennelly-Heaviside layer. It is usually found at a height of 65 or 70 miles, although it may occa-sionally be found as much as 20 miles outside these limits—i.e. at heights ranging from 45 to 90 miles.

A second layer of ionised gas has recently been discovered above this, and is called the F, or Appleton layer, after its discoverer. Its height varies from 90 to 250 miles, and so is even more variable than that of the Kennelly-Heaviside layer. Neither of these two layers reflects all the waves that fall on it, and many waves escape through the Kennelly-Heaviside layer only to be caught and reflected back to earth by the Appleton layer. Indeed, if it were not for this, the Appleton layer could never have been dis-covered.

Other layers have been discovered in the same way, the lowest, known as the D layer, possibly being as little as 25 or 30 miles

above the ground. This layer is specially active in the early morning, trapping long waves, and sending them back to earth. Apart
from this, the majority of waves pass through it quite easily,
although only to be caught and reflected back by one of the upper
layers. When next we listen to a foreign wireless station let us
pause and think out the route by which the radio waves have
brought us the programme. We shall see them leaving the station
and mounting upwards, possibly dodging through the D layer,
and ascending higher until they reach the upper layers, where
they set millions of millions of electrons scurrying about, like
so many goal-keepers in front of miniature goals, trying to prevent the waves from getting through, and kicking such as they
can back to earth. These fall on our aerials, where again they set
electrons scurrying about. If our station is the Daventry National
station with a frequency of 200 kc., every goal-keeper up in the
skies must run backwards and forwards 200,000 times every
second; down on earth the electrons in our aerials run backwards
and forwards just as often, and unless we have a bad contact at
our intake, they also run into and out of our sets, where they
make other electrons jump about inside our valves. And so, as
the result of the varied activities of millions of millions of
electrons, we finally hear the programme.

It may seem strange that there are so many distinct layers of
ionised gas, but we must remember that the atmosphere consists
of a mixture of many kinds of gas, and its different constituents
may be ionised at different heights. Also, ionisation may be
produced by a variety of different agencies, and these may

operate at different heights. The principal agency is perhaps ultra-violet light, which is known to be very potent in ionising molecules of gas. Thus it is significant that all the ionised layers are well above the layer of ozone which shuts off the sun's ultra-violet light.

Quite recently other reflecting layers have been discovered so high up that they must be many miles beyond the top of the atmosphere, and so right out in space. We can tell the height of a reflecting layer by noticing how long it takes for an echo to come back to us from the layer. If, for instance, the echo returns after a delay of a thousandth part of a second, then, since radio waves travel at 186,000 miles a second, we know that the up and down journeys together measure 186 miles, and the layer must be 93 miles high. Now experimenters have recently heard echoes coming back from space after intervals which have ranged from 3 to 30 seconds, shewing that there must be reflecting layers at distances up to nearly three million miles from the earth. Like the nearest layers, these very remote layers probably consist of electrified particles, but these particles cannot be suspended in the atmosphere, because there is none where they are. They are more likely to be electrified particles in transit from the sun to the earth.

For when we go to the sun, we shall find that it is continually shooting out electrified particles, some of which impinge on the earth's atmosphere, after travelling through space for 30 hours or so. It is a general law of electricity that a moving electrified particle is pulled out of its course by a magnet, and we of course know that our earth is a huge magnet. The consequence is that, as these particular particles approach the earth, they no longer travel in straight lines, but are drawn towards the north and south magnetic poles of the earth. Professor Størmer has shewn that, at certain places on their paths, they may be forced to take a very circuitous route indeed, and so be delayed for a long time without making any appreciable progress towards the earth. At such places there must be great accumulations of particles, running round and round for a long time, and it may well be that these accumulations constitute the reflecting layers from which the echoes are heard; the same particles, after they have arrived in the earth's atmosphere, may be responsible for the Polar Lights or Aurora Borealis, which often appear so sensationally in the regions surrounding the north and south magnetic poles of the earth (figs. 32 and 33 on Plate XIV). .

Let us next consider what we can learn about the atmosphere from the passage of sound waves through it. As with radio waves, there are no sound waves falling on to the earth from outer space; there could not be, since sound waves can only travel through an atmosphere, and there is no atmosphere to transmit them in outer space. Thus our study must depend on the noises we ourselves make on earth.

When an explosion or other big noise occurs, waves of sound spread out from it in all directions, much as radio waves do from a wireless station. Those which start in an upwards direction might meet with many fates—the only one to which they cannot be condemned is that of going on for ever in a straight line, because there would be no air for them to travel through. Actually it is found that after travelling to a certain height, they are turned back to the earth by some reflecting layer, much as radio waves are. Our radio sets will often pick up a station 200 miles away while a station 100 miles away is quite inaudible. In the same way, the sound of heavy firing or a big explosion will frequently be heard clearly 200 miles away, although it is quite inaudible at a distance of only 100 miles.

We are all familiar with the children's method of telling the distance of a lightning flash—count seconds between the flash and the thunder, divide by five, and the quotient will be the distance of the flash in miles. The reason for the rule is of course that sound travels through air at about a mile every 5 seconds. Now when the same rule is used to find the distance to a big explosion, it is often found not to work. The sound seems to take too long on its journey, or at any rate longer than it would if it had travelled straight along the ground. It has, in fact, been up to the reflecting layer and down again, and the amount of time by which it has been delayed tells us the height of the layer. Calculation shews that it must be well up in the stratosphere. And now it is easy to surmise what bent the rays back. For we have already seen how the temperature of the stratosphere begins to increase again after a certain height is passed, and it is well known that when sound waves encounter a layer of warmer air, they will be bent back again into the colder air from which they have come.

We can test this property of sound for ourselves, without going up into the stratosphere. Shortly after sunset on a warm autumn evening, mists often form a few feet above the ground, while the air above remains perfectly clear. This shews that the upper layers of air are warmer than the lower, so that in the matter of temperature the two layers—the clear and the misty—form a sort of miniature model of the stratosphere and the troposphere. Under these conditions we shall find that sound travels very distinctly and for great distances along the ground; the waves cannot spread upwards, because every time they try to do so, the upper layer of warm air turns them back again. Similar conditions may often be found over frozen ground at night, and over the surface of a lake at dusk. In each case sound is reflected back to earth just as, at a much greater height and on a far larger scale, it is reflected back by the warmer layers of the stratosphere.

The fall of meteorites, which we shall discuss below (p. 104), provides further evidence that the temperature goes on increasing as we ascend in the stratosphere, and that after being very unpleasantly cold at heights of from 10 to 20 miles, it may become quite comfortable again at a height of about 100 miles.

Fig. 34 (p. 72) gives a diagrammatical view of the different shells or layers of the earth's atmosphere.

We have been so much concerned with the transparency of the earth's atmosphere, that we have hardly remembered that it is not completely transparent, and is often not transparent at all. Living in England, we know only too well how the blue sky may be replaced by clouds, and sometimes by mist or fog.

The Air

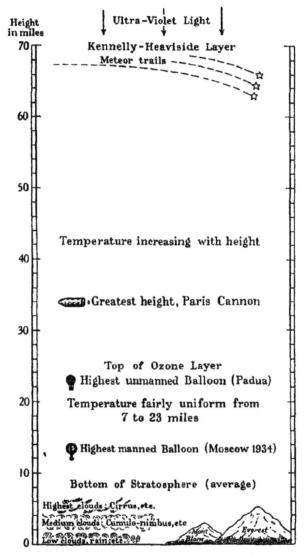

Fig. 34. The results of exploring the atmosphere shewn in diagrammatic form. On the same scale the earth is a sphere 50 feet in diameter.

Moreover, even when clouds, mist and fog are completely absent, there is a sense in which the atmosphere is never entirely transparent. When we arrive on the moon and look upwards, we shall not see a blue sky, but a black one, the reason being that the moon has no atmosphere. In the same way, if the earth's atmosphere were gradually removed, we should see our own sky changing from blue to black. We can see the earlier stages of the process by going up in an aeroplane, thus leaving the greater part of the atmosphere beneath us. Here is the colour of the sky at different heights, as recorded by the observers in the U.S.S.R. balloon "Stratosphere" which went up from Moscow in January 1934:

```
Height  5·27 miles  (8,500 metres)—sky blue.
        6·82   ,,   (11,000   ,,   )— ,,  dark blue.
        8·06   ,,   (13,000   ,,   )— ,,  dark violet.
       13·02   ,,   (21,000   ,,   )— ,,  black-violet.
       13·64   ,,   (22,000   ,,   )— ,,  black-grey.
```

If we could ever go entirely outside the atmosphere, there is no doubt that the sky would look completely black. When we look upwards, we are in effect looking at crowds of particles of air, dust, water vapour, and so forth, every one of which catches some of the sun's rays and scatters them in all directions. Some of these scattered rays enter our eyes, with the result that the sky looks light and not dark to us.

Actually it looks blue, and we may wonder why blue rather than any other colour—for the sun's rays are not specially blue. The explanation is that the sun's light is a blend of waves of different lengths, as we have already seen, and the particles of air, dust, and water vapour do not treat all these different waves in the same way. The waves of blue light are smaller in size than those of red light, while the particles we are now considering are smaller by far than either. Yet because these particles are nearer in size to the waves of blue light than to those of red, they scatter the former waves more effectively, with the result that when we look up at the sky, the scattered rays which enter our eyes are mainly blue, and we say that the sky looks blue. The smaller the particles, the more they scatter the blue light, so that the sky looks bluest of all after heavy rain, when the big dust particles have been washed out; it also looks very blue over the sea and high up a mountain, because here we get away from the more dusty layers. In all these cases only the minute molecules of air are engaged in scattering the light. When the larger dust particles scatter the light we see the familiar haze of a dusty atmosphere.

When we look directly at the sun, the only rays which enter our eyes are those which have not been broken up and scattered. Since the blue rays have suffered more than the red in this respect, it follows that more rays are left of red or reddish colour than of blue, so that the sun looks redder than it really is. If the layer of air or dust between us and the sun is specially thick, as for instance at sunrise or sunset when the sun's rays travel slantwise through the atmosphere, the sun will look even redder than usual. This was observed in a very striking way in the year 1883, when the volcano Krakatoa erupted, and threw out immense clouds of volcanic dust. This dust first shrouded the earth in complete darkness for a distance of 100 miles away from the eruption, and subsequently encircled the world. For the few months during which the earth's atmosphere was permeated with it, the sunrises and sunsets were of an indescribable magnificence.

Particles of water vapour and fog often have a similar effect, so that the sun may look redder when seen through fog. Street lamps shew the same effect, those which are farthest from us looking reddest. Clouds are usually so thick that they blot out the sun's light altogether except near their edges; here we get the proverbial silver, or perhaps golden, lining in the daytime, and the familiar deep red tints at sunset.

The particles of dust, water vapour and fog scatter all the light which tries to pass through them to a greater or less degree, but the red light is scattered less than the blue because its waves are longer. The still longer waves of infra-red radiation are so long that they are hardly scattered at all, so that if our eyes were sensitive to infra-red radiation we should see distant objects through a thick fog just as clearly as we see them through ordinary air.

The camera comes to the rescue to make good the deficiencies of our eyes. We have already noticed that photographic plates are made which are sensitive to infra-red radiation. Fig. 35 (facing p. 76) provides a demonstration of their effectiveness. The two pictures below this (figs. 36 and 37) shew the same landscape photographed simultaneously in ordinary and infra-red radiation. We notice how very clearly the infra-red picture shews distant objects through the intervening haze. When the air is too thick or foggy for our eyes to see distant objects at all, the special infra-red plates can often see them, and the best way of photographing distant objects, fog or no fog, is by using these red plates. Ships have recently been using infra-red photography in the North Atlantic in the hope that it may give them warning of the proximity of icebergs. Infra-red photography also provides a new and very modern proof of the curvature of the earth, since an aeroplane can take photographs of a very distant horizon from such a height that the curvature of the earth is quite distinctly visible.

PLATE XV

Fig 35 An electric flat-iron photographed by its own heat-radiation at a temperature of only 400° Centigrade, at which it emits no visible light

Ilford Co

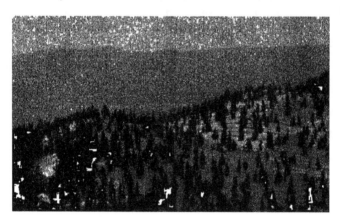

Ilford Co

Fig 36. Bowen Island, British Columbia, photographed by ordinary light from a point 12 miles distant on the mainland.

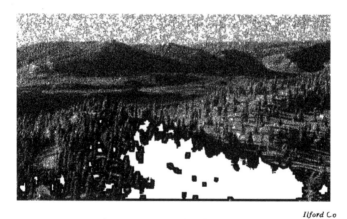

Ilford Co

Fig. 37. The same scene as the above, photographed by infra-red radiation.

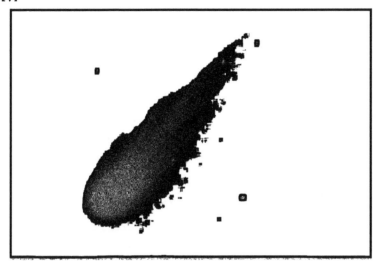

Helwan Observatory

Fig 38. Brooks' Comet (1911)

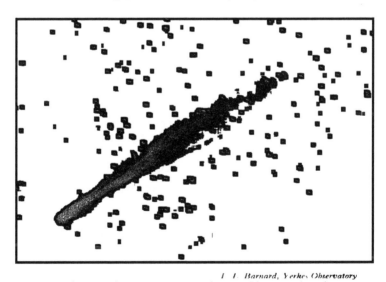

I I Barnard, Yerkes Observatory

Fig 39. Comet III of 1908 The star images are elongated because the telescope was made to follow the motion of the comet, which was moving past the stars.

When the moon is new, it is always near to the sun in the sky; as its size increases, its distance from the sun also increases, until finally, by the time it is full, it is almost exactly in the direction opposite to that of the sun. Because the full moon is opposite the sun in the sky, it is always in the south at midnight.

Whether the moon is new or full or intermediate, the lighted part of its surface is invariably turned towards the sun, and the dark part away. This suggests that the moon emits no light of itself, and only looks bright where it is lighted up by the sun. On somewhat infrequent occasions, the earth comes exactly between the sun and the moon, and so temporarily prevents the sun's light from falling on the moon. We call such an occurrence an eclipse of the moon, and, when it happens, we can see for ourselves that the moon has no light of its own.

There are other, and even rarer occasions, when the moon comes exactly between the sun and the earth—we call such an event an eclipse of the sun. Again the body of the moon, as it passes across the face of the sun, appears as a completely dark screen, and again we have a direct visual proof that the moon emits no light of its own.

Now that all this has been discovered, it sounds as if it must have been simple to discover it. Actually the discovery took a long time. Early man was easily misled by superficial appearances, and so had the most grotesque ideas as to the sizes, motion and physical structure of the sun, moon and stars. For instance, in the sixth century before Christ, the Greek philosopher Anaximander (about 611–546 B.C.) maintained that the sun, moon and stars were simply holes in the firmament, through which fires shone from above. He thought that the phases of the moon resulted from the gradual opening and closing of the moon-hole, while eclipses, both of the sun and moon, occurred when the corresponding holes were completely stopped up.

A few years later we find Anaximenes (about 585–526 B.C.) maintaining that the sun, moon and stars consisted of fire that had risen aloft from the earth. He imagined the sun to be a sort of flat leaf of fire, which floated in the air because of its breadth—rather like a glider or an aeroplane. The moon was something of the same sort, but the stars were of an entirely different nature, being more like fiery studs nailed into the crystal sphere of heaven. As there was nothing in all this to make eclipses, Anaximenes had to suppose that the sky also contained dark bodies "of an earthy nature". Although he did not say so, these presumably made eclipses by coming between our earth and the bright sun and moon.

Next came Xenophanes (born about 570 B.C.) who thought that sun, moon and stars were a succession of clouds of fire sailing across the sky. He believed—as the Egyptians had believed before him—that there was a new sun every day, the sun of the day before having gone so far west as to be invisible; every now and then one of the clouds of fire burned out, and there was an eclipse.

Heraclitus (born about 544 B.C.) thought that the sun, moon and stars were basins or bowls, which collected fiery exhalations from the earth and produced flames from them. The moon-bowl

gradually turned round, and this caused the moon to wax and
wane and go through its well-known cycle of phases. If the
bowls of either the sun or the moon happened to be turned right
away from us there would be an eclipse.

So far no one had got very near to the truth, when suddenly
Anaxagoras (born about 500 B.C.) gave the true explanation of
all the phenomena in one single flash of insight. He said that the
moon was "of an earthy nature, having plains and ravines on
it", and that it derived its light from the sun. He explained how
its phases were the natural consequence of its following the
course of the sun, by which it was illuminated. He also stated
clearly that eclipses of the moon were caused by its passing
within the shadow of the earth when this came directly between
the sun and the moon, and so always occurred at full moon;
while eclipses of the sun were due to the interposition of the
moon between the sun and the earth, and so always occurred at
new moon.

As was not unnatural, the early vague ideas as to the physical
nature of the sun and moon were accompanied by equally vague
ideas as to their sizes and distances. As the sun and moon always
look about the same size in the sky, it is clear that they must
always stay at about the same distances from the earth, but there
was endless difference of opinion as to what these distances were.
Anaximander had maintained that the sun was as big as the earth;
a few years later Heraclitus maintained that it was only a foot in
diameter, while Anaxagoras took an intermediate view, and held
that it was larger than the Peloponnese. The first serious effort to

discover the true facts was made by Aristarchus of Samos (about 310–230 B.C.); he proceeded in the only possible way— by calculations based on actual measurements.

At the moment of half moon we see exactly half the moon's face lighted up by the sun, so that the angle of *EMS* in fig. 40 must be a right angle. If the angle *MES* between the moon and the sun is now measured, all the angles of the triangle *EMS* are

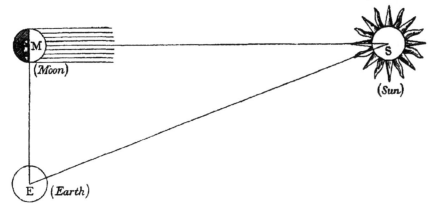

Fig. 40. Diagrammatic representation of the geometrical method by which Aristarchus of Samos tried to measure the distances of the sun and moon.

known, and it is easy to deduce the relative lengths of the sides of this triangle. Aristarchus estimated that the angle *MES* fell short of a right angle by 3 degrees, and deduced that the sun was between eighteen and twenty times as distant as the moon. This was not a good estimate, for, in actual fact, the angle differs from a right angle by less than a twentieth of 3 degrees, and the sun is at about 400 times the distance of the moon.

Aristarchus also had an ingenious means of measuring the distances themselves. At an eclipse of the moon, we see part of the earth's shadow projected onto the face of the moon; it is never more than part, since the whole shadow is far larger than the moon, actually having about four times the diameter of the moon. Aristarchus estimated, however, that the whole shadow was only double the size of the moon, and concluded that the earth itself was double the size of the moon. Having calculated the size of the moon in this way, it was easy to deduce its distance from the angle it subtended in the sky. The moon looks the same size in the sky as a halfpenny held 9 feet away, and for a body 2000 miles in diameter to look as small as this, it must be about 240,000 miles away.

This is a modern calculation; unhappily the measurements of Aristarchus were so faulty throughout that he did not get anywhere near to the true values of the quantities he was trying to evaluate. As we have already seen, he measured the angle *MES* incorrectly, and also took the earth's shadow on the moon to have only double, instead of four times, the diameter of the moon itself. Besides this, he over-estimated the apparent size of the moon in the sky no less than four times, and did not know the dimensions of the earth with any accuracy; some years were yet to pass before Eratosthenes made the surprisingly accurate estimate we have already discussed (see p. 8).

We have seen how the earth is rotating in space, while the so-called "fixed stars" such as Arcturus and Sirius always remain in the same direction in space, and so form a fixed background. The sun and moon appear to move in front of this background, as also do the other objects known as "planets"—from the Greek word πλάνητες, which means wanderers. The five most conspicuous of these—Venus (the morning and evening star), Jupiter, Mars, Saturn, and Mercury—were known before the dawn of recorded astronomy, although it was not always clearly understood that Venus was a single star which appeared alternatively in the morning and evening, or that the same was true of Mercury. The Babylonians, however, appear to have known this, and we find Pythagoras and Parmenides explaining it to the Greeks in the sixth century before Christ. Then, in quite modern times, three more planets were discovered—Uranus in 1781, Neptune in 1846, and Pluto in 1930. Besides these large planets, there are thousands of planets of minute size known as "minor planets" or "asteroids" (p. 143).

To superficial observation the planets appear to wander in very erratic ways. Other astronomical objects move across the sky from east to west with a stately and steady motion, but the planets often fall behind in the procession, and can sometimes be observed moving among the stars from west to east, in what is described as "retrograde" motion. At regular intervals their retrograde motion carries Venus and Mercury backwards across the face of the sun, after which they make a spurt and get in front again; thus the motion of these two planets consists of continual oscillations to and fro about the sun, the westerly swing always being performed much more rapidly than the easterly.

The motions of the planets form so striking a contrast to the orderly motion of the "fixed" stars, that they puzzled the ancients more than a little. The Pythagorean school insisted that the apparent irregularities must be illusory, and that the real motions of the planets must in some way be perfectly even and regular. Geminus wrote that "they could not brook the idea of such disorder in things divine and eternal as that they should move at one time more swiftly, at another time more slowly, and at another time stand still. No one would credit such irregularity even in the case of a steady and orderly man on a journey", while Plato is said to have commended to all earnest students the problem of finding what "uniform and ordered movements" would account for the motions of the planets.

It is a matter of common experience that when an object is performing two distinct motions at once, its actual path in space may be quite complicated, even though each of the two motions is exceedingly simple. If I ride my bicycle along a straight road, the motion of my foot round and round at the end of the pedal-crank is very simple, and so is the motion of the bicycle along the road, and yet my foot moves through space in a very complicated path. The early astronomers tried time after time to explain the complicated paths of the planets across the sky in a similar way.

The first attempt was made in the fourth century before Christ, by Eudoxus of Cnidos (408–355 B.C.). He tried to explain the planetary motions by systems of wheels within wheels—or rather spheres within spheres. These spheres all had the same centre, the earth, but each was pivoted inside the one next outside it, and the spheres all turned in different directions. Each moving object had its own system of spheres, and was supposed to be itself attached to the outermost sphere of the system. Eudoxus found he needed three spheres each for the sun and moon, and four for each of the five planets—twenty-six spheres in all. At a later date Callippus (about 370–300 B.C.) found that even this elaborate system failed to explain the phenomena completely, and added seven more spheres, making thirty-three in all.

The scheme was getting very complicated, but a return to simplicity—and a great step towards the truth—was made almost at the same time by Heraclides of Pontus, whom we have already mentioned as having discovered the rotation of the earth (p. 3). He saw that no complicated systems of wheels or spheres were needed to explain the motions of Venus and Mercury; it was only necessary to suppose that these planets did not revolve around the earth at all, but around the sun like satellites. Then Aristarchus of Samos made an immense step forward by proposing that the earth also revolved around the sun. To quote the description of Archimedes (287–212 B.C.): "Aristarchus of Samos brought out a book consisting of certain hypotheses, in which the premises lead to the conclusion that the universe is many times greater than that now so called. His hypotheses are that the fixed stars and the sun remain motionless, that the earth revolves about the sun in the circumference of a circle, the sun lying in the middle of the orbit, and that the sphere of the fixed stars, situated about the same centre as the sun, is so great that the circle in which he supposes the earth to revolve bears the same proportion to the distance of the fixed stars as the centre of the sphere bears to its surface".

Such views as these were not popular in the days of ancient Greece—or in any other days. Man has never liked being told that his home in space is not the hub of the universe, as he has so often fondly imagined, but a mere speck circling round another speck, on so minute a scale that the whole is only like a point in the vast sphere of the universe. And so we read in Plutarch how Cleanthes thought that Aristarchus ought to be indicted for the impiety of putting into motion the Hearth of the Universe—i.e. the earth. Aristarchus had told men a truth which they found unpalatable, but they easily found other astronomers who were only too ready to tell them everything that they wanted.

Throughout nearly two thousand years after Aristarchus, the most favoured explanation of the motion of the planets was one of cycles and epicycles—not the wheels within wheels of Eudoxus, but rather wheels upon wheels. Heraclides had supposed that Mercury and Venus wheeled round the sun while the sun itself wheeled round the earth. It was soon found that an extension of this scheme would explain the motion of all the astronomical bodies. Thus the earth was still made the centre of the universe in spite of Aristarchus; A wheeled round the earth, B wheeled round A, and C around B, and so on—like the house that Jack built—until a point on the rim of the last wheel was found to reproduce exactly the observed motion of a planet.

About A.D. 150 Ptolemy of Alexandria put this theory of cycles and epicycles in a form which held almost unchallenged sway throughout the intellectual darkness of the Middle Ages. Here and there a doubter may have been found to express his doubts, but no serious challenge occurred until A.D. 1543, when the Polish monk Copernicus proposed replacing the whole system of Ptolemy by one very like that which Aristarchus of Samos had propounded eighteen hundred years earlier. In brief, he supposed that the sun stood still, while the earth and the other five planets all revolved around it. Two-thirds of a century later, the telescope of Galileo established the truth of his views.

Such views proved to be no more popular in mediaeval Europe than they had been in ancient Greece, and Copernicus had, with great worldly prudence, withheld the publication of his book until he lay on his death-bed; Galileo, less endowed with prudence of this particular kind, boldly proclaimed what he believed to be the truth and found himself in frequent trouble with the ecclesiastical authorities throughout the rest of his life.

As Aristarchus and Copernicus had in effect proclaimed, the planets only appear to move irregularly because we on earth view the scene from a non-central position; we are like the spectators at a theatre who cannot see the play in its proper setting because they are too far to the right or left of the stage. The sun provides the proper central position from which to view the planetary motions, and an observer who established himself on the sun would see each planet repeating the same almost circular path over and over again with the utmost regularity. He would also see that the paths of the planets were all very

nearly in the same plane, a plane which is inclined at a small angle
of about 7 degrees to the sun's equator.

Just as such an observer, from his position on the sun, would see
our home, the earth, moving round him in a circular path across
the sky, so from our position on earth we see his home, the sun,
moving in a circular path across the sky. This apparent path of
the sun across the sky is called the "ecliptic", and, as all the other
planets move nearly in the same plane as the earth, we see these
also moving across the sky in almost the same path as the sun.
The three nearest planets—Venus, Mars and Mercury—may at
times be as much as 9, 7 and 5 degrees respectively away from it,
but none of the other five planets ever go as much as 3 degrees
away. Thus the paths of the sun and planets all lie within a quite
narrow track across the sky. This narrow track was known to the
ancient Egyptians and Babylonians, and also, probably through
the Babylonians, to the ancient Greeks. It is called the "Zodiac".

These early races of course regarded the stars merely as points
of light, but they could hardly help noticing that these points of
light fell rather naturally into the groups which we call "constellations". They named these after animals, heroes of legend, or
familiar objects—sometimes from a supposed resemblance which
was often rather fanciful, but more often for no apparent reason
at all. The Babylonians divided the Zodiac into twelve equal
parts, and placed one constellation in each. All of these were
originally named after animals, and all but one still are. The word
Zodiac means "animal circle", and the twelve constellations
were originally supposed to be the houses of animals which the

sun visited in turn, one each month, as it moved across the sky. For astronomical reasons, it is usual to start the list with the month of April, or, more precisely, with the spring equinox. We can remember the twelve constellations in their proper order by a jingle written by Dr Watts, the hymn writer:

> The Ram, the Bull, the Heavenly Twins,
> And next the Crab, the Lion shines,
>   The Virgin and the Scales,
> The Scorpion, Archer, and He-Goat,
> The Man that holds the watering-pot,
>   The Fish, with shining tails.

The Greeks and Egyptians had very similar names for many of the constellations of the Zodiac, but the Chinese Zodiac is named after twelve quite different animals. In place of our Ram, Bull, Twins and Crab, they have Dog, Cock, Ape, Ram, and so on.

The remainder of the sky has also been divided into constellations, some of which are mentioned by very ancient writers. Orion and the Great Bear are mentioned both in Homer and in the Book of Job, while the Little Bear was described by Thales in the seventh century before Christ. Many of the constellations also are common to many languages and peoples. The Orion constellation, for instance, is often associated with a hunter or hero, and the Taurus constellation with a fierce animal.

All the constellations which could be seen from ancient Greece were drawn on a globe by the astronomer Eudoxus, a pupil of Plato, in the fourth century before Christ, and subsequently described in verse by Aratus. Most of them are associated in some

way or other with the legends or fairy tales of long ago—either
of ancient Greece or of some still earlier civilisation. Thus we
read of Helice and Cynosura, the Great Bear and the Little Bear,
the latter being a hunter who was changed into a bear so that he
should not kill his mother, whom Juno had already changed into
a bear out of jealousy; or again of Hercules (whom Aratus describes merely as "The kneeling man") and the dragon; or—
best story of all, a real thriller—of Perseus arriving in the nick of
time to rescue Andromeda who was chained to a rock in the sea
while Cetus, the sea-monster, was coming to devour her. He
made Cetus look at the Medusa's head, which turned everyone to
stone who saw it, but escaped this fate himself by looking at
it in a mirror. I have heard it suggested that our more modern
nursery rhyme, which describes the cow jumping over the
moon, was inspired by the sight of the moon moving through,
or perhaps under, the constellation Taurus. The little dog who
laughed to see such fun would no doubt be Canis Minor, the next
constellation. There is also a dish (Crater) in the sky to run away
with the spoon.

The Greeks were not great travellers, so that there were parts
of the sky south of the equator which they did not see at all, and
so could not divide into constellations. It was a pity, for the
moderns who named the constellations in this part of the sky did
not always maintain the dignity and simplicity of the older
names. We find such constellations appearing as the Printer's
Workshop, the Painter's Easel, the Engraver's Pen, the Chemical
Furnace, and, even more ridiculous, the Honours of Frederick,

the Harp of the Georges, the Oak-tree of Charles I. Even more recently a French astronomer, Lalande, tried to insert a cat into heaven. He wrote: "I love cats; I adore cats; I may be pardoned for placing one in the sky after sixty years of arduous labours". But it has since disappeared, perhaps because it did not enjoy the society of its neighbours, Canis Major, Canis Minor, and Canes Venatici.

As Greece lies about 40 degrees north of the equator, the parts of the sky which the ancient Greeks could not see would be those which lay within 40 degrees of the South Pole. We might then reasonably expect that all the constellations with modern names would lie inside a circle 40 degrees in radius, having the South Pole as centre.

Broadly speaking, we find that they all lie within a circle of 40 degrees radius, but its centre is not the South Pole. The reason for this is both interesting and informative.

The earth spins in space like a spinning top, but its axis does not always point in the same direction. The bulge round the earth's equator is continually being pulled by the sun's gravitational pull, and as this pull twists the earth's axis round in space, the earth top wobbles, rather as the ordinary schoolboy's top does when it is "dying".

It is found that the earth's axis wobbles round in a complete small circle once every 26,000 years. At the present moment the axis points to the tip of the tail of the Little Bear, but 4000 years ago it pointed to the Bear's left ear, and 5000 years ago to the tip of its nose. And 13,000 years ago the whole Little

Fig. 41 *a.* The constellations north of the ecliptic. The North Pole is not at the centre of the figure, but is approximately where the small (dotted) circle cuts across the tail of the Little Bear. As the earth wobbles about in space, the north pole of the earth's axis moves round and round this small circle, at the rate of one revolution every 26,000 years.

Fig. 41 *b*. The constellations south of the ecliptic. The constellations of the Zodiac (see p. 89) lie along the circumference of either this or of fig. 41 *a*, opposite.

Bear was well down in the northern sky, while the earth's axis pointed near Vega, which is now well down in the sky. Because the spinning top on which we live is rolling about in space, the inhabitants of Greece must have seen different parts of the sky at different epochs—just as, when we live on a rolling ship, we see different sights through the porthole of our cabin. This goes some way towards explaining why many southern constellations, such as the Centaur, have Greek names; those parts of the sky are not visible from Greece now, but they were 4000 years ago, when people believed in Centaurs.

The constellations which Aratus mentioned in his poem are not even those which the Greeks were able to see at the time of Aratus; they are, broadly speaking, those which had been visible from the latitude of Greece about 2500 years earlier, or about 2800 years before Christ. Thus it seems likely that Aratus merely described constellations which had been named in the first instance by people who resided in the same latitude as Greece, at the period of about 2800 B.C. This points very strongly to the Babylonians, especially as there is other evidence that some at least of the principal constellations had been known to the Babylonians at an even earlier date.

The constellations owe their familiar outlines, and sometimes their names as well, to the brightest of their stars, but they also contain a great number of fainter stars, many of which we can only just see with our unaided eyes, and a still greater number which cannot be seen at all without a telescope.

With average eyesight and good conditions, the human eye

can just see the light of a single candle at about 6 miles distance. If we dim the light, or move to a greater distance than 6 miles from it, we shall not see a fainter light, but no light at all. Thus the light we receive from a single candle 6 miles away forms what we may call the "threshold of vision".

Let us take the amount of light we receive from a single candle 6 miles away as our unit of brightness, so that the faintest stars we can see with our unaided eyes have a brightness of exactly one unit. On this scale the star which looks brightest of all, Sirius, is found to have a brightness of 1080 units—in other words, it looks as bright as a 1080 candle-power lamp would at a distance of 6 miles—while the second brightest, Canopus, far down in the southern sky, has only 550 units. These two stars are quite out-standing; their nearest competitors are a succession of stars with about 200 units of brightness each—Vega with 220, Capella with 205, Arcturus with 200, α Centauri and Procyon with 180 each, and so on. There are only about twenty stars in the whole sky which shine with 100 units of brightness or more. After these come about 200 more which shine with between 100 and 10 units of brightness, and then about 4500 more shining with between 10 and 1 units. This completes the list of stars that we can see with average unaided eyesight—the stars that have more than one unit of brightness. We see that there are only about 4720 in the whole sky—not only in the part we can see, but in the part below the horizon as well. Not more than about half of this number will be above the horizon at any one moment, and even of these a fair proportion are likely to be hidden by the mist or clouds near the horizon. On the whole, we shall be lucky if we see as many as 2000 at any one moment with average eyesight, although naturally people with specially good eyesight may expect to see more.

Most people find it hard to believe that the number is so small; if they are asked to guess how many stars they can see, they usually put the number far too high—except, of course, those who have read books on astronomy, and so know the answer already.

There is another guessing game in which the victim is invited to guess the greatest number of threepenny pieces that can be laid flat on a half-crown without overlapping. The answer is one, but most people are sure they can put two on—until they have tried. Here is a similar question: How many visible stars does the full moon conceal? In other words, if the moon suddenly became transparent, so that we could see through it, how many stars should we be likely to see lying behind it with our unaided eyes? The answer is none at all, which again most people find hard to believe.

The sun and moon are so bright that most people over-estimate their sizes enormously. Each of them takes a whole day to move round the sky, and we can easily verify that each takes only about 2 minutes to cross the length of its own diameter—in other words, the whole sun or moon moves past any fixed point in 2 minutes. This shews that it would take 720 suns or moons placed in contact side by side to make a circle round the sky. From this we can calculate that if we had to wallpaper the whole

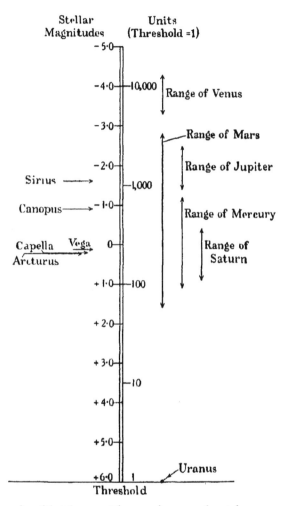

Fig. 42. The scale of brightness. The numbers on the right represent multiples of our unit of brightness, the threshold of vision being unity. The numbers on the left represent "stellar magnitudes", by which the astronomer measures brightness in a more technical manner. The relation between the two measures is obtained by comparing numbers on the two sides of the central line.

The brightnesses of a few stars are shewn on the left-hand side of the diagram, and those of the planets on the right.

sky with suns and moons, we should need about 200,000 of either. This is rather more than forty-two for each visible star, so that there is a chance of less than one in forty of there being a visible star behind the moon.

As soon as we bring telescopic power to our aid, the number of stars we can see naturally increases by leaps and bounds. The primary function of a telescope is to collect the waves of light which fall on a large area—the object glass or mirror of the telescope. It then throws these waves into our eyes much as an ear trumpet collects sound waves and throws them into our ears. The diameter of the human eye is only a fifth of an inch, so that a telescope of 1 inch diameter will collect twenty-five times as much light as an unaided eye, and enables us to see stars whose brightness is anything above a twenty-fifth of a unit. There are about 225,000 such stars, so that even a 1-inch telescope will shew us 220,000 stars more than we can see without it—nearly fifty new stars for every old one. The great 100-inch telescope at Mount Wilson will shew us stars of only about a three-millionth part of a unit, the total number of these being perhaps 1500 million. Yet even this vast number, as we shall see later, is only about one per cent. of the total number of stars.

In spite of the immense number of stars, their total light, as we know, is not overpowering. Indeed the total light we receive from all the stars in the sky other than the sun is only about 100,000 units—rather less than a hundred-millionth part of the light of the sun. It is equal to the light of a single candle 100 feet away.

The stars shine by their own light, but the planets only because they are lighted up by the sun. Naturally, then, a planet sends out enormously less light than a star, but its nearness may often make up for the feebleness of its light. Indeed it occasionally more than makes up, so that a planet may often appear the brightest object in the whole sky.

The inexperienced observer will not always be able to distinguish planets from stars by a mere superficial glance at the sky, although it may help him to remember that a planet can never wander by more than a very short distance from the ecliptic, the central line of the Zodiac. The brightest planets, Venus, Mars and Jupiter, can frequently be identified from their mere brightness (see fig. 42, p. 97). Venus, when it can be seen at all, is always the brightest object in the sky, but Mars and Jupiter may be either brighter or fainter than the brightest star, Sirius.

Most of the stars shine with a steady light, and as they stay at the same distance from us, their brightness does not vary. The brightness of the planets, on the other hand, varies for two reasons. As they move round the sun, their distance from us continually varies, as does also the fraction of their surface which we see illuminated. These changes are most marked in the case of our nearest neighbour Venus, its illuminated surface and apparent diameter varying in the way shewn in fig. 43 (p. 100). It is clear that Venus cannot look brightest when it is nearest to us, because then we only see a thin crescent of it illuminated—like the new moon; neither does it appear brightest when its full surface is illuminated, because it is then so far away from us that its surface looks very small. It is brightest when it is in an intermediate position in which its appearance is as shewn in fig. 43 (*c*). It then shines with 13,000 units of brightness, and so looks twelve times as bright as Sirius. When Mars and Jupiter are at their

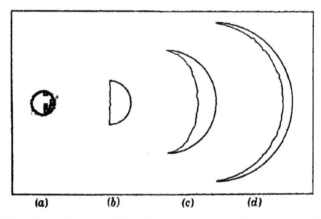

(a)   (b)   (c)   (d)

Fig. 43. *The phases of Venus:* (*a*) when at its greatest distance from the earth—a circle 9½ seconds in diameter; (*b*) when farthest from the sun in the sky—a semicircle 18 seconds in diameter; (*c*) when its brilliancy is greatest—a crescent 40 seconds in diameter; (*d*) when of the largest diameter at which it can be seen—a crescent 62 seconds in diameter.

As the planet passes even nearer to the sun, its diameter further increases to 63 or 64 seconds, but it is then so much immersed in the sun's glare as to be invisible.

brightest, they shine with 3300 and 2500 units respectively, so that they can both be considerably brighter than Sirius, but the other planets cannot rival the brightest stars, Mercury having only 760 units at its best, and Saturn only 360 units.

The full moon has a brightness of 26 million units, and so is two thousand times as bright as Venus at its best, while the sun, in the full light of day, shines with 12,300,000,000,000 units, and so is nearly half a million times as bright as the full moon.

It may seem surprising that such lights as these do not blind us, when our eyes are so sensitive that they can see as little as a single unit of light. Actually our salvation lies in the fact that the effect on our eyes does not depend so much on the number of units of brightness, as on the number of digits in this number—or, more strictly, on what the mathematician calls the logarithm of the number. The effect on our eyes is not given by

Sun 12,300,000,000,000; Venus 13,000; Sirius 1080; Faintest star 1;

but rather by

Sun 14; Venus 5; Sirius 4; Faintest star 1;

and now the sun does not look so overpowering.

Although the positions and brightnesses of the planets are continually changing, the sky looks much alike night after night, so that its changes usually cause no surprise. Yet occasionally far more exciting apparitions are seen than the orderly procession of sun, moon, planets and stars. Foremost among these less usual sights are comets and shooting-stars. To the untutored savage comets must look like stars which have gone crazy, and rush about the heavens with their hair streaming out behind; indeed, early writers used to refer to all comets indiscriminately as "the hairy star", as though there could be but one such remarkable object in the sky. Shooting-stars look—not only to the savage but to everyone else as well—very like stars which have lost their foothold in heaven and have fallen to earth. We shall discuss the physical constitution of these objects later; at present we are only concerned with their appearance and movements in the sky.

Comets move round the sun like the planets, but in very different paths. A planet moves nearly in a circle, and so stays always at about the same distance from the sun, but a comet usually moves in a very elongated orbit, and its striking appearance is usually restricted to the few weeks or months in which it is nearest to the sun. During this time the radiation of the sun causes the comet to eject a long tail, which invariably points away from the sun. Typical examples of comets are shewn in figs. 38 and 39 (facing p. 77).

Before their true nature was understood, comets were regarded as portents of evil, and, oddly enough, many of the most conspicuous appearances of comets seem to have coincided with, or perhaps just anticipated, important events in history. Homer (*Iliad*, 19) writes of:

> The red star, that from his flaming hair,
> Shakes down diseases, pestilence, and war.

It was not until Newton had explained the motions of comets, shewing that they obeyed the same laws of motion, and were guided by the same gravitational pull, as the planets, that they ceased to be regarded in this sinister light.

Shooting-stars may make even more sensational displays in the sky. These often come singly, but often also in showers. Occasionally on looking up on a clear night we may see dozens of these, sometimes myriads, darting through the sky like huge

PLATE XVI

W I Gordon

Fig 44 The huge Hoba meteorite, estimated to weigh 60 tons.

W I Gordon

Fig. 45. A pile of iron meteorites collected in the
Gibeon district of south-west Africa.

Royal Geographical Society

Fig 46 Meteor Crater near Cañon Diablo, Arizona. The view is an air photograph, taken from the north-west, and looking almost exactly along the direction in which the meteor is believed to have been moving when it struck the earth.

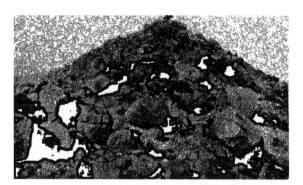

Royal Geographical Society

Fig 47 The aerial photograph shewn in fig 46 does not convey an adequate idea of the immense size of the Meteor Crater Fig 47 shews the highest point on its rim, Barringer Point, with a man on horseback at the summit to give an indication of the scale.

fireflies. The early Chinese and Japanese seem to have been greatly affected by shooting-stars and kept careful records in which they are described as falling like snow, or heavy rain, or leaves from the trees in autumn. Here is a description of a Korean meteor shower in A.D. 1519, which has been unearthed by Y. Iba of the Kobe Observatory:

Some shooting through the sky like arrows gone astray, some rampantly ascending like red dragons, some bursting like fire-balls, some curling like bended bows, while others looked like bifurcated bodkins, and transformed themselves into many motley shapes and appearances.

Actually these objects have no right to be described as stars at all. They are not immense bodies, millions of miles away in space, but tiny fragments of hard rocky or metallic substance, most of them so small that we could hold hundreds, perhaps even thousands, in one hand. And they are quite near home—in our own atmosphere, in fact.

Tiny pellets of hard material are continually travelling through outer space; millions of them strike the earth's atmosphere every day, travelling at hundreds of times the speed of a rifle bullet. When they first enter the atmosphere, the friction of the air causes them to become first hot, then very hot, then red hot, and finally white hot; it is at this stage that they look like stars. They become so hot that, after a brief life of only a few seconds' duration, they disintegrate into gas and dust and disappear from sight.

It may seem surprising that so small an object as a shooting star can look as bright as a real star such as Sirius or Arcturus, but we must remember two things. First, the shooting-star is much nearer—it is playing to a much smaller audience, only a few miles of earth instead of millions of millions of miles in space. Second, it shines for a far smaller time—only a few seconds, whereas the real stars shine for thousands of millions of years at least.

Of the same nature as these small bodies are the larger bodies called meteors. When these dash through the air, they often appear far brighter than any star, and may light up the whole landscape; we then describe them as fire-balls. Sometimes their outer surfaces become so hot that they crack and burst—just as a cold glass may burst if hot water is suddenly poured over its surface—and they often make loud, and even terrifying, reports as they do so. For instance, a Japanese record of date A.D. 1533 tells us that "stars dazzlingly scintillated all over the sky, and shot down to the land and sea, breaking into pieces like stones and giving out tremendous clangours, so that there were fears lest the earth might be knocked about, and the Kingdom decay, and the whole populace lamented awfully in dismay".

Such displays were frequently regarded as evidence of the displeasure of the gods, and often resulted in kings and nations altering their ways of living. Livy tells how a fall of meteors in 650 B.C. led to a nine days' solemn festival in the hope of propitiating the angry gods, and the Japanese records tell of many occasions on which the whole nation set about mending its ways after the supposed admonition by a fall of meteors. The diary of Columbus tells us how, even after his sailors had seen tropical birds, and so must have known they were near the long looked-for land, "they saw a meteor fall from heaven, which made them very sad".

The little shooting-stars invariably dissolve into vapour before they reach the earth, but generally speaking the larger meteors do not; they usually fall to earth, and are then described as meteorites. The smaller may lie about in deserts or in farmlands until they are discovered and removed to museums or to laboratories for analysis. The majority prove to be mere stones or masses of crystalline rock, but a few consist of iron, sometimes mixed with rock or stone and sometimes with nickel and cobalt. Figs. 44 and 45 on Plate XVII (facing p. 102) shew the largest of known meteorites and a pile of smaller iron meteorites.

Even larger meteorites may bury themselves in the earth, and often make great holes or craters where they fall. Plate XVIII (facing p. 103) shews two views of a large hole in the earth in Arizona, which is known as the "Meteor Crater"; it is oval in shape, with a circumference of 3 miles, and a depth of 570 feet. It is conjectured that this must have been formed by the fall of a huge meteor about 500 feet across, and weighing perhaps 14,000,000 tons. Plate XIX (facing p. 106) shews views of a group of similar but smaller craters, known as the Henbury Craters, in Central Australia; the largest has dimensions of about 220 yards by 120 yards, and is about 50 feet deep. We can hardly doubt that these craters were formed by the fall of meteors, since masses of meteoric iron have been found in all of them.

These craters were made by meteors which fell in pre historic times, so that we know nothing of the circumstances of their falling. But one which fell in Siberia in 1908 shewed how much havoc a meteor can make when it buries itself in the earth, and then explodes. Trees were singed and blown down for distances of more than 30 miles from the centre—thousands of square miles devastated by the fall of a single meteor. It is difficult even to imagine what the surrounding country must have looked like after the fall of far greater meteors in Arizona and Central Australia.

Fig 48 The Henbury Craters, Central Australia General
        view of the main crater, taken from the air

Fig 49. Another view shewing the interior of the same crater.

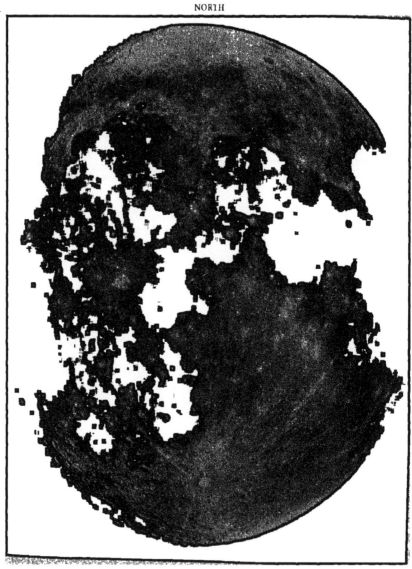

Puiseux, Paris Observatory

SOUTH

Fig. 50 The moon 12½ days old, photographed with a 24-inch telescope. Parts of the moon's surface are shewn in greater detail on Plates XXI–XXIV.

CHAPTER IV

THE MOON

We know that the moon always looks about the same size in the sky and from this we can conclude that it is always at about the same distance from the earth. And we can measure the distance in the same way as we measure the distance of an inaccessible mountain peak, or the height of an aeroplane.

When an aeroplane is up in the air, people who are standing at different points must look in different directions to see it. If it is directly overhead for one man, it will not be directly overhead for another man a mile away, and its height can be calculated simply by noticing how far its position appears to be out of the vertical for the second man. Using this method, astronomers find that the distance of the moon varies between the limits of 221,462 miles and 252,710 miles, the average distance being 238,857 miles. Thus, in round numbers, we may think of the moon as being a quarter of a million miles away.

At such a distance, we can hardly expect to see much detail with our unaided eyes. Indeed, as we watch the moon sailing through the night sky, we can detect nothing on its surface beyond a variety of light and dark patches, which, with a bit of imagination, we can make into the man in the moon with his bundle of sticks, or an old woman reading a book, or—as the Chinese prefer to think—a jumping hare. Naturally no sane people have ever thought these creatures actually resided in the moon, but in past ages many people thought that the moon was a huge mirror which merely reflected the features of the earth, so that what appeared to be light and dark patches on its surface were simply the reflections of our own lands and seas; others thought that the dark patches were objects suspended in space between ourselves and the moon. We have seen how Anaxagoras, who first explained the phases and eclipses of the moon (p. 80), declared that "the moon is of an earthy nature, having plains and ravines on it".

As soon as we look at the moon through a telescope, or even through a pair of field-glasses, the mystery of its structure is solved, as Galileo found when he turned his newly-made telescope onto it in 1609. He announced at once that the moon was a world like our own, having its own seas and mountains. For a long time the dark patches were believed to be seas of real water, and they were named accordingly. For instance, the three largest "seas" in the upper half of Plate XX (facing p. 107) are named in succession, running from left to right:

"Mare Imbrium"—The Sea of Showers.
"Mare Serenitatis"—The Sea of Serenity.
"Mare Tranquillitatis"—The Sea of Tranquillity.

Yet we know now that these cannot be seas of real water, since we never see the glitter of sunshine reflected from them as we so often do from a distant lake in a landscape on earth. As the moon moves and turns about in space, the sun's rays fall on it from all directions in succession, but a shining reflection of the sun has never yet been seen, and we now believe that the so-called seas

are really dry deserts. We can understand why Tranquillity and
Serenity were chosen as appropriate names for the supposed
lunar oceans—nothing was ever seen to happen there. The Sea
of Showers was a less appropriate choice—in fact a mere effort
of imagination; perhaps the early astronomers felt the need for
variety.

Not only is there no water on the moon, but there is also no air
or atmosphere of any kind, unless in quite inappreciable amount.
This is shewn very clearly when the moon eclipses the sun by
passing in front of it. Just at the end of the eclipse a moment
comes, the last moment of darkness, when the vividly bright sun
is just about to emerge from behind the dark moon—the sun is,
so to speak, about to rise from behind the lunar mountains. Now
if the moon possessed any atmosphere at all, the sun's coming
would be heralded by tints of dawn, just as it is when the sun
rises behind mountains on earth. But in the actual occurrence
nothing is seen until the sun bursts forth in full brilliance.

A large modern telescope enables us to see a great deal of detail
in the scenery of the moon, and to photograph even more. For
it can readily be transformed into a huge camera, and the driving
clock of the telescope will turn this to follow any part of the
moon, or any other object in the sky we please, so that a
photographic plate can be exposed for any length of time without fear of blurring.

Plate XX (facing p. 107) shews the moon, when nearly full,
photographed through the 24-inch telescope of the Paris Observatory. To make this look the same size as the actual moon in

the sky, we must set the picture up at a distance of 50 feet from where we stand. If we now illuminate the picture, we shall be able to pick out the man in the moon, the old woman, the hare, and so forth. Then if we walk towards the picture, we shall see all these imaginary inhabitants gradually dissolving into plains and mountain ranges.

The four photographs shewn in Plates XXI–XXIV were taken with a still larger telescope—the great 100-inch telescope at Mount Wilson—and shew various details of lunar scenery.

We know how objects on earth cast very long shadows at sunrise and sunset, but shorter shadows when the sun is high up in the sky. It is the same, of course, on the moon, and the heights of the lunar mountains can be estimated from the lengths of the shadows they cast at various times of the lunar day. Although the moon has only a quarter of the diameter of the earth, its mountains are found to be rather higher on the average than those of the earth, a great number being more than 15,000 feet in height, while many are far more precipitous.

So far we have merely been looking at the moon from a distance. Let us now charter a rocket to take us there, so that we can actually walk on its surface.

Our rocket must be shot off at a high speed—6·93 miles a second at least—for if it starts at any lesser speed it will merely fall back to earth, like the shot from an ordinary gun. If it starts with a speed of exactly 6·93 miles a second, it will just get clear of the earth's gravitational pull, but after it has got clear, it will have no appreciable speed left to carry us on our

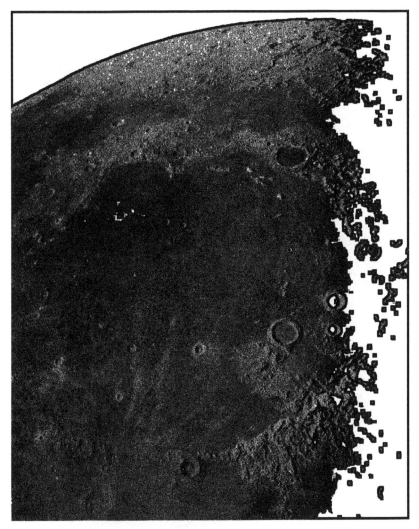

Mt Wilson Observatory

Fig 51 A part of the northern half of the moon, which can easily be identified on Plate XX. The big "sea" which occupies the centre of the plate is the Mare Imbrium, the range of mountains which bounds it to the south-east is the Apennines At the southern extremity of this is the big and deep crater named Eratosthenes, while still lower down and to the left is the even larger crater Copernicus, which is shewn on a larger scale on Plate XXIV (facing p. 113)

PLATE XXII

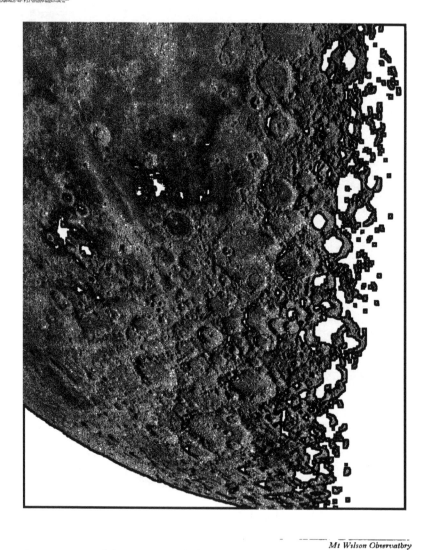

Mt Wilson Observatory

Fig 52 A part of the southern half of the moon The "sea" which is about half an inch from the left-hand edge will readily be identified with the most southerly sea on Plate XX, it is the Mare Humorum To the right of it lies the larger Mare Nubium.

journey. Let us start it with a speed of 7 miles a second, then it will still have a speed of 1 mile a second left after it has got clear of the earth's pull, and we shall reach the moon in a little over 2 days.

We only take a few seconds to pass through the earth's atmosphere, which is relatively hardly thicker than the thin skin of a plum or a peach. As we pass through this, we gradually leave beneath us all the particles of air, dust, water vapour and so on, which scatter the sun's light and make the sky look blue. As the number of these particles decreases, we see the sky assuming in turn the colours already described (p. 73)—blue, dark blue, dark violet and black-grey. Finally we leave the earth's atmosphere beneath us and see the sky become jet black, except for the sun, moon and stars. These look brighter than they did from the earth, and also bluer because none of the blue light has been subtracted from them to make a blue sky. And the stars no longer twinkle at us as they did on earth because there is no atmosphere to disturb the even flow of their light. They seem now to stab our eyes with sharp steely needles of light. If we look back at our earth, we shall see about half of its surface shrouded in mists, clouds and showers. But in front, the whole surface of the moon shines out perfectly clear; it has no atmosphere to scatter the sun's light, and no fogs and rains to obscure the illumination of its surface.

Naturally this clearness persists after we have arrived on the moon's surface, and far exceeds anything we have ever experienced on earth. We have seen how our atmosphere is the

cause of the soft tones that add so much to a terrestrial landscape—
the oranges and reds of sunrise and sunset, the purples and greens
of twilight, the blue sky of full day, the purple haze of the distance. Here on the moon there is no atmosphere to break up
the sun's rays into their different colours and distribute them—
the blue to the sky, the red to the dawn, and so on. There are only
two colours—sunshine and shadow, white and black; everything
in the sunshine is white, everything else black. We feel as though
we were in a cinema studio lighted only by one terribly powerful
light—the sun. A valley stays utterly dark until the moment
when the sun rises over the surrounding mountains; then full
day comes, with all the suddenness of turning on an electric
light.

It is clear that if we want to step out of our rocket and walk
about on the moon, we must bring our own air with us; we shall
need an oxygen apparatus, such as the climbers on Mount
Everest had. We may perhaps think that the weight of this will
make walking or climbing very arduous, but as soon as we set
foot on the soil of the moon, we shall find that the contrary is the
case. The moon contains less than an eightieth part of the substance of the earth, and so exerts a gravitational pull which is
much smaller than the earth's—in fact it is only about a sixth as
great. For this reason, we find we can carry extraordinary
weights without fatigue, and as our bodies seem to weigh almost
nothing, we can jump to great heights. We feel so athletic that
we may even try to break our own jumping records. It ought
not to be difficult to break both our own and everybody else's;

PLATE XXIII

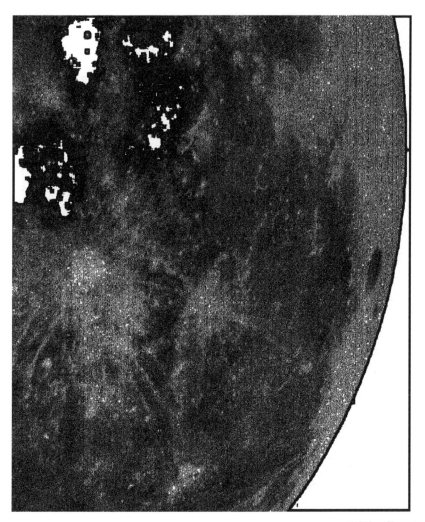

Mt Wilson Observatory

Fig 53. This is part of the edge of the moon which will readily be identified rather more than half-way down on Plate XX The "sea" at the left-hand edge of the plate is the Mare Nectaris, to the right of this is the Mare Foecunditatis, and above it a deep bay of the Mare Tranquillitatis

PLATE XXIV

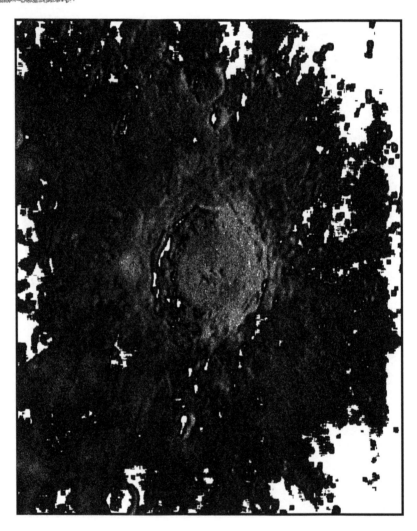

Mt Wilson Observatory

Fig 54 Detail of the crater Copernicus which is 46 miles in diameter It is easily
found at the bottom edge of Plate XXI Fig 58 (facing p 116) may help us visualise
how the crater would appear to a traveller on the moon.

a good high jumper ought to jump about 36 feet, and the long jump of a fair athlete ought to be at least 120 feet. If we feel inspired to play cricket, the ball will simply soar off our bat, so that if it is not to be entirely a batsman's game, the pitch and field must each be six times the size they are on earth. Unfortunately, all this will make the game six times as slow as on earth, and perhaps cricket played six times as slowly as on earth, would not be much of a game after all.

If we fire a gun, our shot will travel a terrific distance before falling back to earth—or rather to moon. We remember the big guns which fired shells nearly eighty miles in the Great War; if similar guns were mounted on the moon, their projectiles would go right off into space and never return. We shall not want to start setting big guns up on the moon, but we can produce the same effect with something much simpler—a breath of air from our breathing apparatus.

For we know that ordinary air consists of tiny particles, called molecules, which are incessantly jumping about—some quite slowly, the majority at about the speed of a rifle bullet, and a few at far higher speeds. Some move faster than any projectile which has ever been fired from a gun.

We had to start our rocket from earth with a speed of about 7 miles a second, in order that it might overcome the earth's gravitational pull; with any lower speed it would have merely fallen back to earth like a cricket ball. And a projectile of any other kind needs precisely the same speed if it is to get clear of the earth. Now it is only at very rare intervals that molecules of air attain a speed of 7 miles a second, so that they seldom jump right off the earth into space—this is why the earth retains its atmosphere. On the other hand, a projectile only needs a speed of $1\frac{1}{2}$ miles a second to jump entirely clear of the moon, and molecules of ordinary air frequently attain speeds as high as this. We see at once that an atmosphere of air could not survive on the moon for long, since each molecule would jump off into space the moment it attained this critical speed of $1\frac{1}{2}$ miles a second.

Just because there is no atmosphere on the moon there can be no seas, rivers or water of any kind. We are accustomed to think of water as a liquid which does not boil away until it reaches a temperature of 212 degrees, but if ever we picnic high up on a mountain, we find out our mistake; we soon discover that water boils more easily and at a lower temperature there than on the plain below. The reason is that there is less weight of air to keep the molecules of the liquid pressed down, and so prevent them flying off by evaporation. If there were no air-pressure at all, the water would evaporate no matter how low its temperature, and this is precisely what would happen on the moon. Clearly then we shall find no water on the moon; we must take drinking water with us, and it will not be well to pour it out and leave it standing; if we do it will have disappeared by the time we want to drink it—its molecules will have danced off, one by one, into space.

Knowing that there is neither air nor water on the moon, we shall hardly expect to find men or animals, trees or flowers. And

PLATE XXV

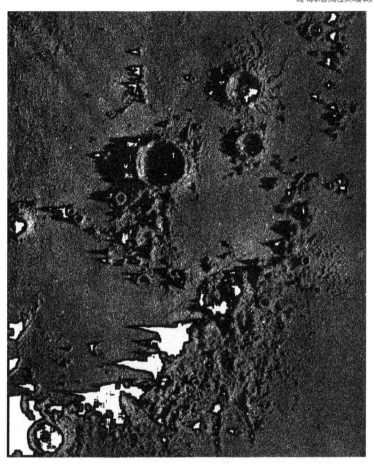

James Nasmyth

Fig. 55. Slanting across the lower half of the plate are the Lunar Apennines; above them is the large crater Archimedes. This scene is easily recognised on the right of the photograph shewn in Plate XXI.

PLATE XXVI

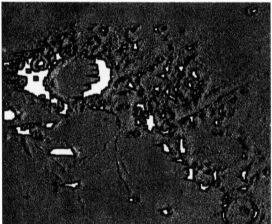

J Nasmyth

Fig. 56   The large crater to the left is Plato, while the furrow to the right and slightly lower down is known as the "Valley of the Alps"   Both are easily recognised in the upper right-hand part of the photograph shewn in Plate XXI

J Nasmyth

Fig. 57.   The isolated mountain which can be seen to the south of the crater Plato, either in fig 56 above or on Plate XXI, is known as Pico   It rises directly from the plain to a height of 8000 feet, and would probably look somewhat like this to a pedestrian walking on the moon.

in actual fact, the moon has been observed night after night and year after year for centuries, and no one has ever found any trace of forests, vegetation or life of any kind. No changes are detected beyond the alternations of light and of dark, of heat and or cold, as the sun rises and sets over the arid landscapes. The moon is a dead world—just a vast reflector poised in space, like a great mirror reflecting the sun's beams down onto us.

Let us now step out of our rocket, and survey the lunar scenery. I cannot shew you detailed pictures of what we shall see, but I can do the next best thing. About 50 years ago, the engineer James Nasmyth made calculations of the heights of a great number of the mountains in the moon, both large and small, and constructed a model to shew the results he had obtained. Fig. 55 on Plate XXV (facing p. 114) illustrates a small part of the model, which is easily recognised again in Plate XXI. Fig. 56 shews another region which also appears in Plate XXI. The small isolated mountain to the right is named Pico, and Nasmyth's drawing of this is reproduced in fig. 57. Figs. 58 and 59 on Plate XXVII (facing p. 116) are imaginative drawings of scenery of yet other kinds.

It is natural to wonder why the scenery of the moon is so different from that of the earth. Is the moon formed of different stuff from our earth, or was it formed out of similar stuff but in a different way, or can the whole difference be traced to a difference of physical conditions?

We have already seen how our terrestrial mountains, volcanoes, craters, etc., were formed. In brief, the earth started life as

8-2

a ball of very hot gas, which shrank and cooled and then liquefied, so that it finally became rather like a sponge filled with drops of liquid and bubbles of gas. Then the sponge shrank still further, and the bubbles were squeezed out to form oceans and atmosphere. A solid crust formed, and as this too shrank, it wrinkled up to form mountain-ranges, such as our Himalayas and Alps. These may originally have been five or ten times as high as at present, but have been flattened and smoothed out by rain, snow and frost.

Now it seems probable that the lunar mountains also were formed in the first instance as wrinkles on the cooling moon. But the gases and water vapour which were expelled from the interior could not stay encircling the moon in the form of atmosphere and seas; their molecules would simply soar off into space. Thus the factors which have smoothed the outlines of our terrestrial mountains have been lacking on the moon from the very outset, and the lunar mountains have remained perfectly clear-cut in shape.

Yet something must have happened on the moon to give the mountains those sharply-cut outlines; they are broken rocks, and something must have broken them. Indeed, observers have occasionally seen what they have believed to be clouds of dust such as might result from falls of rock. As there is neither rain nor ice on the moon to break up its rocks, there must be something else at work. If we take a walk on the moon, we may soon find out what this is.

We have seen how hard fragments of rocky or metallic sub-

PLATE XXVII

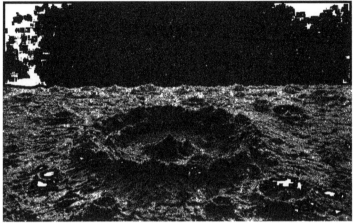

Nasmyth and Carpenter, " The Moon"

Fig 58. This does not attempt to represent any particular crater on the moon, but is typical of the kind of scenery which occurs in regions where the whole lunar surface is mottled with craters of all sizes.

Nasmyth and Carpenter, " The Moon"

Fig, 59 A typical landscape in a mountainous part of the moon, at a moment when the sun is just being eclipsed by the earth. The bright ring round the earth is produced by the earth's atmosphere; the band of light is the zodiacal light.

stance are continually bombarding the earth's atmosphere from outer space. The smaller fragments live a brief, but very vivid and brilliant, life as shooting-stars, and evaporate harmlessly into dust before they reach the surface of the earth, but we have seen that the larger fragments may do a great deal of damage.

Similar bodies must of course be continually bombarding the moon, but here they find no atmosphere to check their fall and to dissipate the majority into dust before they can do any harm. Big and little meteors alike strike the surface of the moon with exactly the motion with which they have previously been moving through space—like a rain of small bullets and big cannon-balls. I have read a great many stories of travels on the surface of the moon, but their writers all forgot that the explorers would be under a continuous hail of fire from these objects. The experience might not be altogether amusing. At a moderate computation, more than a million shooting-stars and meteors must strike the surface of the moon every day, their speeds averaging perhaps 30 miles a second, which is about 100 times the speed of a rifle bullet. And such a speed as this makes them formidable, even though their size may not. At a speed of 30 miles a second, a tiny pellet of matter has as much energy—and also as much capacity for doing damage—as a motor-car moving at 30 miles an hour, while a half-pound meteor has the same energy as the Royal Scot rushing along at 70 miles an hour; there would not be much left of a house if such a meteor fell on it. Clearly we terrestrials owe a good deal to our atmosphere for saving us from this sort of adventure. And we can see that the impact of meteors provides a quite sufficient explanation of any clouds of dust or falls of rock that may have been observed on the moon.

It has sometimes been conjectured that falls of meteors may also have produced the ring-shaped formations which form so conspicuous a feature of a lunar landscape. Such falls may have produced some of the smaller craters, but cannot have produced them all. For if they had all been produced in this way, we should expect them all to look rather like the meteor craters we find on earth. Actually they differ in many respects. The largest of the ring-formations on the moon are far larger than any meteor craters known on earth, and are also far more regular in shape. Meteor craters, being produced by the impact of meteors at all sorts of oblique angles, may shew any degree of elongation or irregularity, but nearly all the lunar craters are circular in shape, and this seems to shew that they were produced by something acting inside the moon rather than by something coming from outside. A great number have central elevations like the vent holes of terrestrial craters, and this suggests that we must attribute them to the same kind of volcanic action—in brief they seem to be the craters of extinct volcanoes.

This and other evidence makes it likely that the surface of the moon consists in the main of a mass of volcanoes and their outpourings of lava and volcanic ash. On the earth, the combined influences of air, rain and frost disintegrate volcanic outpourings and transform them into soil, which ultimately sustains vegetation and life, but on the moon there is nothing to act on the products of volcanic eruption, and change their quality, so that these are likely to stand for ever as lava and ash.

It is possible to test this conjecture scientifically. In fig. 59 (facing p. 116) the artist has depicted an eclipse of the sun. Let us imagine such an event occurring during our visit to the moon. What shall we find?

We must expect our foremost sensation to be one of extreme cold. Those of us who have been present at an eclipse of the sun on earth know that it can get fairly cold when the sun's light is suddenly shut off, but the earth has warmth stored in its atmosphere and soil which saves us from being completely frozen. Here on the moon there is no atmosphere to store up warmth, and we cannot expect much from the soil, since volcanic ash is an exceedingly poor conductor of heat, being just about as poor as the asbestos which the plumber packs round hot-water pipes to prevent the heat escaping. Even if the moon's interior stays reasonably warm, its warmth will do us little good, since we shall be on the wrong side of a thick asbestos-like screen. Thus, when the sun's light and heat are shut off, we must expect the more than tropical heat of the full sun to give place to a cold more intense than anything known on earth.

And this is what happens. In a factory we may occasionally see a workman pointing an instrument, known as a pyrometer, towards some point of an oven or fire to discover its temperature. In precisely the same way, in an observatory an astronomer may occasionally point a telescope furnished with a thermocouple towards a star or a point on the moon's surface to discover its temperature. In this way the changes of temperature on the moon's surface can be followed through the various stages of an eclipse. The changes are found to be quite sensational, both in amount and rapidity. As the earth's shadow passes across the face of the moon, and covers any particular spot in darkness, the temperature at that spot may be observed to fall from about 200 degrees Fahrenheit to about 150 degrees below zero within a few minutes.

Such a rapid fall of temperature suggests at once that but little of the heat stored in the moon's interior comes up to its surface, which of course means that the moon's surface layers must be bad conductors of heat. Actual calculations shew that they must have just about the same feeble conducting capacity as volcanic ash.

Equally violent changes of temperature occur at the ordinary rising and setting of the sun, although of course not with the same startling rapidity. The temperature may be as low as 250 degrees below zero Fahrenheit before sunrise and may have risen to more than 200 degrees Fahrenheit, or about the ordinary temperature of boiling water, by the time the sun is directly overhead. Through all these changes, the blanket of volcanic ash keeps the interior of the moon at a fairly uniform temperature; if we dig only about an inch down we shall come to a steady temperature somewhere near to that of melting ice.

There are still other ways of discovering what the moon is made of. Judging by its appearance, people have guessed that it is made of all kinds of substances—ice, snow, rocks, silver and even green cheese. We cannot, however, tell what an object is made of by merely looking at it; a great many substances look alike that are really very different in structure, as, for example, diamonds and paste, or real pearls and false. We may do better if we look at our object in a number of different coloured lights in succession, for substances that look alike in one light will often look very different in another.

Now the spectroscope enables us to do just this; it sorts out the different colours of light and lets us use them separately. We can, so to speak, let each colour of light tell its own story—by itself and undisturbed by the others. In a police court, the magistrate insists on the witnesses speaking separately; the policeman describes the accident, and tells how he arrested the man who is charged with furious driving, the people who saw the accident say in turn what they saw, the owner of the car tells his story, and so on. It would be hard to arrive at the truth if they all shouted at once. Now each different colour of light that we receive from objects out in space has got its own story to tell of the nature of the objects from which it comes, and the spectroscope enables us to hear the different stories one at a time.

Although two different substances may conceivably look similar in a few isolated colours of light, they are sure to give different records for some colour or other. Thus, if two substances behave in the same way for all colours, and so give identical records throughout the whole range of the spectrum, we may be reasonably sure that they are of identical material.

In Plate XV (p. 76), we have already seen a landscape photographed in infra-red and in ordinary light. We notice at once that different kinds of objects give very different records, and this shews that they are made of different substances. When, however, the moon is photographed in the same way, all its various parts are found to give similar records, not merely in these two colours of light, but in all other colours of light as well. We conclude that all parts of the moon's surface are made of much the same substance. Further, if we can find any substance in the laboratory which again gives the same record in all colours of light, we shall suspect that it is of similar structure to the moon's surface.

There is a further and more technical method of study which leads to even more definite results. Light can not only be broken up into waves of different lengths (i.e. into different colours), but also into waves which vibrate in different directions. When we play on a violin string with a bow, the string vibrates in more or less the direction in which it is dragged by the bow, and this is parallel to the body of the instrument. But if we pluck the string by hand, it vibrates in the direction in which we pluck it, and this may be perpendicular to the former direction. The string gives out the same note as before, but its vibrations take place in a different direction.

Now when light is reflected by any substance, the direction of its vibrations is turned round in space, and the extent to which it is turned depends very largely on the nature of the substance. Thus we can to some extent identify substances by the extent to which they turn the plane of vibration of light.

Before we finally conclude that the moon is made of any particular substance, it is important to test whether the substance in question turns the plane of vibration about in the right way. The test is a very stringent one, for we can test not only for each colour of light, but also for each colour reflected at every possible angle.

Now volcanic dust or ash passes this test quite triumphantly, and indeed reproduces the actual record of the moon's surface in every respect, except for one small spot close to the crater Aristarchus. This looks black in ultra-violet light, but hardly shews in ordinary light. Its record can be matched by volcanic rock stained by a thin deposit of sulphur spread lightly over it —and sulphur is a common ingredient in the outpourings of volcanoes on earth.

Thus, taking all the evidence together, it seems very probable that the moon's surface consists of volcanic ash; it looks like volcanic ash not only in light of one colour but in lights of all colours; it rotates the plane of vibration of light in the same way as volcanic ash, not only for one colour but for all; it behaves like volcanic ash in its very low capacity for conducting heat; and, finally, it lies at the feet of what are almost certainly volcanoes.

CHAPTER V

THE PLANETS

There are nine planets circling round the sun, of which of course the earth is one. Of the other eight, five have been known from pre-historic times, while the remaining three—the three farthest from the sun—are comparatively recent discoveries.

The row of models exhibited in fig. 60 shew how greatly these nine planets differ in size. Those which are nearest to, and farthest away from, the sun are the smallest, while the middle members, Jupiter and Saturn, are the largest. Jupiter, the central member, is largest of all, with a diameter of nearly 90,000 miles, and a volume 1300 times that of the earth. Jupiter stands in the same proportion to the earth as a football to a marble, while on the same scale Mars would be hardly larger than a pea.

If we wish to complete our model by placing the objects shewn in fig. 60 at their proper distances, the nearest planet, Mercury, must describe an orbit which is not quite circular, but is such that, even at its nearest approach to the sun, the planet would be 20 feet away. The earth must keep at a distance of 50 feet from the sun, while Pluto, the farthest planet of all, must describe an orbit nearly half a mile in radius.

We see that the solar system consists mainly of empty space, and yet the emptiness of the solar system is as nothing compared to the emptiness of space itself. For if we wish to continue constructing our model on the same scale, we must place the

PLATE XXVIII

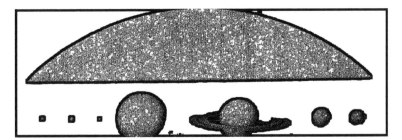

Carnegie Institution

Fig 60 The relative sizes of the sun (above) and the planets (below) of the solar system The planets are arranged from left to right in order of their distance from the sun Mercury, Venus, Earth (with moon), Mars, Jupiter, Saturn (with rings), Uranus, Neptune, Pluto

W H Wright, Lick Observatory

Fig 61   Venus in ultra-violet light (left) and in infra-red light (right)

W H Wright, Lick Observatory

Fig 62 Mars in ultra-violet light (left) and in infra-red light (centre) The composite picture on the right shews that the ultra-violet image is larger than the infra-red, the difference in size resulting from the thickness of the Martian atmosphere.

nearest fixed star nearly three thousand miles away—somewhere near New York. We see that space is very empty.

The nine planets all move round the sun in the same direction, and as we have seen, very nearly in the same plane, thus making a sort of regular "one-way" traffic. All except Mercury, Venus and Pluto—the planets nearest to and farthest from the sun—have one or more satellites, the giant central planets Jupiter and Saturn being exceptionally rich with at least nine each, and probably more, for Dr Jeffers of Lick Observatory has recently discovered a tiny object which moves with Jupiter and is believed to be a minute tenth satellite of Jupiter only a few miles in diameter.

With unimportant exceptions, all the satellites move round their planets in the same direction in which the planets themselves move round the sun, and approximately in the same plane.

Besides the planets and their satellites, there are thousands of bodies known as minor planets or asteroids, which again move round the sun in this same direction; 1264 such bodies were known at the end of 1933. There are also a large number of comets, again moving round the sun in this same direction. The "rule of the road" is the same throughout the solar system. How is this rule enforced, and how is the traffic regulated and kept going?

If the planets were left entirely to themselves, each would move steadily onward in a straight line, and soon lose itself in the depths of space. We, on earth, should find ourselves running off at a rate of 19 miles a second into the petrifying cold of outer space.

Yet the history book we explored in the first chapter tells us that the earth has been at much the same distance from the sun for many millions of years past. We can only conclude that something is holding the earth in, and preventing it running off into space, just as when we see a horse running round and round a groom in a field, we conclude that something is holding the horse in.

This "something" is of course the sun, and its hold is what we describe as the force of gravitation. You probably all remember how—at least according to the story—Isaac Newton watched an apple fall to the ground, and reflected that if the earth attracted to itself objects which were near its surface, such as apples, it must also attract to itself bodies far out in space, such as the moon. He did not expect that the earth would pull on a distant object as strongly as on a near one, but thought the pull would probably weaken according to the inverse square of the distance—the law according to which apparent brightness of an object, as well as many other quantities in nature, are observed to fall off.

If so, it would of course be possible to calculate the earth's pull on the moon. The moon is sixty times as far from the centre of the earth as we are, so that the earth's pull ought to be 3600 times as strong here as out where the moon is. Here it causes objects to fall 16 feet in a second towards the earth. There—if Newton were right—it would cause objects, including the moon itself, to fall only 1/3600th part of this—rather more than a twentieth of an inch—in a second towards the earth. Small though this is, it is exactly the fall needed to keep the moon in its orbit, and prevent it flying off into space. Although the moon travels at a speed of nearly 2300 miles an hour—forty times the speed of an express train—yet the continual repetition, second after second, of this small fall earthwards, results in its being no farther from the earth now than it was a thousand years ago.

Just as the earth's pull keeps the moon moving in an approximately circular path round the earth, so the sun's pull keeps the earth and the other planets moving in circular, or nearly circular, paths round the sun. Each planet may be compared to a weight whirled round our hand at the end of a string. Our hand is the sun, and the pull in the string is the sun's gravitational pull. The faster we whirl a weight round, the greater the pull in the string by which we hold it. Now observation shews that the nearest planets are being whirled round the sun far more rapidly than the outermost, so that the sun's pull on the nearer planets must be far more intense than on the outer. This fits in with Newton's law that the intensity of the force of gravitation falls off according to the law of the inverse square of the distance. Indeed it is this law that determines the speeds and distances at which the planets move; they adjust their speeds and distances so that the force of gravitation on each planet is of exactly the intensity needed to keep the planet moving round and round in its orbit.

Naturally, then, Mercury completes its journey round the sun in far less time than Pluto; actually it moves completely round the sun in about 3 months, and so alternates between being a morning and an evening star about eight times every year, while Pluto, which takes a thousand times as long—250 years—to travel round the sun, stays in the same part of the sky for year after year. The other planets naturally move round the sun in times which are intermediate between these extremes—Venus in a little over 7 months, the earth as we know in a year, Mars in rather less than 2 years, Jupiter in nearly 12 years, Saturn in 29½ years, and so on.

The sun pours out heat and light in all directions like a fire; the planets are like a number of sentries walking round and round the fire. The nearest man may be uncomfortably hot, while the farthest may be very cold, unless he has private supplies of heat to keep him warm independently of the warmth he receives from the fire.

If a planet has no supplies of heat stored up in its interior, it will radiate out into space precisely the amount of heat that it receives from the sun. This amount is easily calculated, but the amount that the planet radiates out into space depends on the temperature of its surface—the hotter the surface, the greater the amount of radiation. A planet with no internal supplies of heat will assume the temperature at which there is an exact balance between its receipts and expenditure of radiation. If it were spinning very rapidly round its axis, its whole surface would stay uniformly at this temperature—just like a leg of mutton which is being turned round and round in front of a fire. Actually most of the planets rotate rather slowly on their axes, and that side of a planet which has been facing the sun for a long time must obviously be a good deal hotter than the opposite side which has been kept in the dark. The consequence is that the night temperatures of the planets are generally well below the day temperatures, and the temperature at any point on a planet's surface is not steady, but continually fluctuates about an average.

The amount of heat that the earth receives from the sun is easily calculated, and to radiate this amount of heat away into space the earth would have to be at an average temperature of about 40 degrees Fahrenheit, which is only just above the freezing-point of water—such at least would be the case if the earth were a hard black sphere without any atmosphere. Small adjustments must be made on account of the earth's atmosphere and the quality of its surface, and after all this has been allowed for, we find that the calculated average temperature of the earth is somewhat lower than the mean temperature actually observed, which is about 57 degrees Fahrenheit. This shews that the earth does not obtain all its heat from the sun, but must have slight internal supplies of heat—probably the radioactive substances in its crust, of which we made the acquaintance in our first chapter.

We can calculate in the same way the average temperatures which the other planets would assume if they were warmed solely by the heat of the sun; these range from about 343 degrees Fahrenheit for Mercury to about 380 degrees below zero Fahrenheit for Pluto. On the whole these calculated temperatures are fairly close to the temperatures actually measured with a thermocouple, shewing that all the planets, like the earth, have but little internal heat of their own, and derive their warmth almost entirely from the sun's radiation.

The temperature of Mercury, the planet nearest to the sun, is of special interest. Calculation shews that, if Mercury rotated rapidly, its surface would be at a uniform temperature of 343 degrees Fahrenheit. The more slowly it rotated, the more of course its temperature would fluctuate about the average. In the extreme case in which it always turned the same face to the sun— as the moon does to the earth—one face would be permanently far above 343 degrees, and the other permanently far below; calculation shews that a point which was at the centre of the hot face, and so for ever directly under the sun, would be at a temperature of about 675 degrees. Now the observed temperature of a point directly under the sun is very near indeed to this, and this proves that the planet always turns the same face to the sun— in other words, Mercury has one face on which it is always day and another on which it is always night. The day face, with its perpetual temperature of about 675 degrees, is far too hot for water to exist on it in liquid form. It is also too hot for any atmosphere to be retained. For Mercury only contains about a twenty-fifth part as much substance as the earth, and its gravitational pull is so much less than that of the earth, that a molecule or any other projectile would fly right off into space as soon as its speed reached about $2\frac{1}{4}$ miles a second. Molecules would frequently attain these speeds in the grilling heat of the hot face, so that if Mercury ever had an atmosphere, this must long ago have flown off into space. We obtain a direct visual proof of this when Mercury crosses in front of the sun. It looks like a perfectly sharp black disc, and, just as with the moon, this shews that Mercury possesses either no atmosphere at all, or so little as not appreciably to refract the sun's rays.

The planet is usually so near the sun as to be completely lost in its glare, and even at the best of times it is exceedingly difficult to see anything of its surface. Yet certain permanent markings can be discerned on it, rather like those that we see, with incomparably greater clearness, on the face of the moon. A study of the light reflected from the planet suggests that its surface may be very similar to that of the moon, possibly a rough surface of volcanic ash or dust.

Venus comes next after Mercury in order of nearness to the sun, and has the special interest of being the planet which is most like our earth. In many respects it is a sort of twin sister to the earth. It has almost the same diameter—3870 miles as against the earth's 3960—but its substance is rather less closely packed, with an average density only 4·86 times that of water, as against 5·52 times for the earth. As a consequence of this, Venus has 19 per cent. less total substance than the earth, and the pull of gravity at its surface is 15 per cent. less than that at the surface of the earth. A molecule or other projectile will leave the surface of Venus and fly off into space as soon as its speed reaches 6·3 miles a second, as against the 6·93 miles a second needed on earth.

So far the two planets are clearly very similar; what differences there are result mainly from Venus being much nearer to the sun than the earth is. Calculation shews that Venus ought to have an average temperature about 90 degrees Fahrenheit higher than that of the earth. Even with such a temperature, however, water could still exist in liquid form, and the planet could retain an atmosphere, so that we should expect to find seas and rivers, atmosphere and clouds, storms and rain on Venus, much as on earth.

Certainly our expectations with respect to atmosphere and clouds are fully confirmed. On the very rare occasions when Venus passes in front of the sun, it presents a very different appearance from those of Mercury and the moon, which have no atmosphere. At the moments when Venus encroaches upon and leaves the bright face of the sun, we do not see it as a sharply defined and clearly outlined black disc, but as a dark disc rimmed with pearly light, the light being produced by the refraction of the sun's rays as they pass through the atmosphere of the planet. A general study shews that the planet is completely encased in clouds—clouds which are so thick and ever-present that it is impossible to see through them, even though we take advantage of the cloud-penetrating properties of infra-red light.

Fig. 61 (facing p. 124) shews pictures of Venus taken in infra-red and ultra-violet light, and there is clearly no essential difference in quality between the two. A few dark markings can be seen in the ultra-violet picture, but these are not permanent, and are probably only specially dark patches of cloud or fog. If they were features of the solid surface of the planet they would be most conspicuous in the infra-red picture. Thus we must regretfully abandon all hope of ever seeing any sort of solid surface beyond the clouds.

It is not difficult to understand why Venus should be encased in clouds and fog in this way, since its higher temperature must obviously result in more water being kept in a state of evaporation than on earth. But, whatever the reason may be, the clouds are there, and in such profusion as to make all study of the lower reaches of the planet's atmosphere impossible; we can, so to speak, only study the stratosphere of Venus, the region above the clouds and fogs.

We studied the composition of the earth's stratosphere by examining sunlight which had travelled through it. We found that this light had been robbed of certain of its constituent wavelengths, and deduced the presence of ozone in the stratosphere.

We can use a similar method for Venus. We see its clouds by light which has passed through the stratosphere of Venus twice on its journey from the sun to us—once in passing down to the clouds, and a second time in coming back again, after reflection, from the clouds to our eyes. Again, when this light is compared with light which has come directly from the sun to us, certain wave-lengths are found to be missing. As the loss can only have occurred in the stratosphere of Venus, we can deduce the composition of this stratosphere.

It is at once found to be different in composition from the stratosphere of the earth. It contains no appreciable amount of water vapour, but perhaps this is hardly surprising, as there is not much in the stratosphere of the earth. There is a more marked difference in the fact that Venus hardly contains any appreciable amount of oxygen in its stratosphere. To assess the importance of this, we must remember that most chemical substances shew a great hunger to combine with oxygen, as we see in the familiar processes of rusting, corrosion and combustion. Indeed this oxygen hunger is so intense that it is perhaps rather surprising that any oxygen at all is left in the atmosphere of the earth. That there is any left is probably due to the circumstance that the earth's oxygen is continually being replenished by the vegetation which so profusely covers the surface of the earth. This acts as a vast oxygen factory, and the fact that no oxygen can be found on Venus may very possibly mean that there is no vegetation on Venus to supply it.

To imagine the physical conditions prevailing on the surface of Venus, we must take a very long step away from those we left behind us on the moon and on Mercury, in the direction of conditions now prevailing on our own earth. On the surfaces of the moon and Mercury we should probably find arid and rocky deserts, which are roasted in the presence of the sun or frozen in its absence, and are uniformly undisturbed by either wind or rain. There may also be rocky deserts on Venus, but they cannot be arid, and some changes at least must occur. If Venus turned on its axis as rapidly as our earth, we might confidently expect to find such familiar and home-like phenomena as trade-winds, dry and rainy seasons, and so forth. We have, however, already seen that Mercury always turns the same face to the sun, and it seems highly probable that Venus either does the same or else alternates its faces with extreme slowness. In other words, Venus may have a day face and a night face like Mercury, or may have days and nights which are like our own except for being of extreme length. In either case there may be very little either of wind or rain, merely a perpetually damp and hot climate.

The surface of Venus may well be like that of the earth was in those far-distant days before life had appeared to change the appearance of its surface and the composition of its atmosphere. As we travel backwards in time, we must come to an epoch when the earth was substantially hotter than now, either because it still had appreciable stores of internal heat, or because the sun was itself hotter than now, and so provided a more abundant supply of radiation. It may be that the Venus of to-day provides a picture of the earth of those days, and that the Venus of the future is destined to repeat in some measure the history of our own earth. Even if vegetation is still lacking on Venus, it may appear in due course, and by supplying the atmosphere with oxygen, may open the road to higher and higher forms of life. Yet we know so little of the nature and meaning of life that all such thoughts are at best the wildest of guesses. For aught we know life may be destined to take very different forms on Venus, or may never appear at all. We simply do not know and have no right to guess.

Still travelling out into space and increasing our distance from the sun, we pass by our own earth, which we have already studied sufficiently, and come next to Mars. If Venus is a twin sister of the earth, Mars is the earth's little brother. If Venus is a warmer edition of the earth, Mars is a much colder edition. If Venus suggests a picture of what the earth may have been in the remote past, Mars suggests what the earth may possibly be in the remote future.

Mars cannot compare with Venus or the earth in bulk and substance, having only a little more than half the diameter, and only a little more than a tenth of the substance, of the earth. Also its substance is less densely packed than that of either the earth or Venus, so that its gravitational pull is quite small. With the same effort we shall be able to jump three times as far, or three times as high, as on earth—as against six times on the moon. A molecule or other projectile only needs a speed of 3·1 miles a second to jump off into space, as against the 6·9 needed on earth. If Mars were as near to the sun as Mercury is, the molecules of its atmosphere would attain this speed quite frequently, so that most, or all, of them would probably have disappeared by now. But the greater distance of Mars from the sun has saved it from this fate, and a considerable thickness of atmosphere is still left. Fig. 62 (facing p. 124) shews two photographs of Mars, taken at Lick Observatory in ultra-violet and infra-red light respectively. When the halves of the two photographs are joined together we obtain the third picture on the extreme right of fig. 62, and see at once that the ultra-violet picture is distinctly larger than the infra-red. The difference of course represents the thickness of the Martian atmosphere.

As with Venus, the light by which we see the surface of Mars has passed twice through the whole thickness of the Martian atmosphere, so that again we might expect that certain wavelengths would be missing from the light, and that from the missing wave-lengths we could deduce the constitution of the Martian atmosphere. But when the light is analysed it is hard to find that anything is missing. The Mount Wilson astronomers have a very powerful equipment at their disposal, and have specially searched for evidence of either oxygen or water vapour in the atmosphere of Mars. They can find no evidence of oxygen, and consider that there cannot, at the most, be as much as a thousandth part as much oxygen per square mile of surface as there is in the atmosphere of the earth.

They find no direct evidence of water vapour in the atmosphere either, although it has often been thought that there is a certain amount of circumstantial evidence that water vapour is present. Mars has its alternation of hot and cold seasons as we have, and it is noticed that certain features of its surface change regularly with the seasons. White caps, for instance, appear round the poles in the cold season and disappear in the warm season. It has often been conjectured that these may be ice or snow—perhaps clouds of icy particles in the air, or perhaps fields of snow on the ground—although it is of course also possible that the snow may be merely carbon dioxide or some substance quite other than frozen water vapour.

It is also noticed that dark patches appear regularly in the Martian spring, and fade away again in the autumn—mainly in the tropical regions and southern hemisphere. It was at first thought that these were real seas of water, but this is now considered improbable. For one thing, they vary too much and too rapidly in colour; one, for instance, was observed to change from blue-green to chocolate-brown and back again within a very few months. They also resemble the supposed seas on the moon in never reflecting the sunlight, as sheets of real water would do. At one time astronomers thought they might be forests, or masses of vegetation. Since then the surface of Mars has been examined in the same way as the surface of the moon, and appears to be of somewhat similar composition—possibly volcanic lava or some such substance. Thus the dark patches may be produced by showers of rain wetting a dead dry surface like that of the moon.

If we are planning to take our rocket to Mars, it is clear that we must again take air and water with us. We must also be prepared for an exceedingly inhospitable climate, and we may as well know the worst before we start.

Mars has days and seasons very like our own. It takes 24 hours and 37 minutes to turn on its axis, so that its day is slightly longer than ours. And as this axis is tilted at an angle of 25° 10', as against the earth's angle of 23° 27', we must expect to find the Martian seasons rather more pronounced than ours on earth; there will be a greater difference between summer and winter. On top of this, however, there is a further cause of variation in the climate on Mars.

The earth's path round the sun is very nearly circular—not quite, since the earth's distance from the sun is 3 per cent. less in December than in June. We inhabitants of the northern hemisphere are closest to the sun at our mid-winter, while people in the southern hemisphere are closest at their mid-summer. Thus the small variations in our distance from the sun go to lessen the

difference between summer and winter in the northern hemisphere, but accentuate it in the southern hemisphere. As a
consequence, we must go to the South Pole rather than to the
North for extremes of climate.

Nevertheless, the earth's distance from the sun does not vary
enough to produce any great effect on our climate. It is different
with Mars, whose path is nothing like so circular as that of the
earth. Our distance from the sun varies by less than 3 million
miles, but that of Mars varies by more than 26 million miles.
Thus, when Mars approaches the sun, the climate of the whole
planet becomes appreciably warmer; as it recedes, the whole
planet gets colder. These alternations of general coldness and
general warmth are of course superposed on top of the ordinary
Martian seasons. The maximum of general warmth, the time
when Mars is nearest the sun, occurs shortly before mid-summer
in the southern hemisphere, so that on Mars, as on earth, we must
go to the southern hemisphere for extremes of climate. Further-
more, the extremes will be far more marked than with us.

Now if we are planning to land our rocket on Mars, we may as
well take advantage of what warmth there is—even so, we shall
soon find there is little enough. Let us then arrange to arrive
when Mars is nearest the sun—i.e. at the middle of the period of
general warmth—and to land slightly south of the equator at
mid-day. Here we may find a temperature as high as 60 degrees
Fahrenheit. But if we cherish any hopes that we have come to a
fine, warm climate, they will be dispelled as evening closes in.
For there are neither clouds nor atmosphere enough to retain

the planet's warmth, so that it will get cold with great rapidity as soon as the direct radiation of the sun diminishes—just as on a terrestrial desert, only far more so. It is likely to freeze before sunset, and may well fall to 40 degrees below zero before the sun reappears on the scene.

This is the very best climate Mars can offer. If we travel to the poles we must expect to encounter temperatures of more than 100 degrees Fahrenheit below zero, while if we wait until the planet is at its farthest from the sun, the temperature will be still further reduced all over the planet, and we may be unable to find any spot on the planet's surface at which the temperature is above the freezing-point.

We have already seen that the surface of Mars is probably rather like that of the moon, so that when we step out of our rocket we must expect the general nature of the scenery to be rather like what we found on the moon. We can hardly expect to find any vegetation, at any rate of the kind we know on earth, since it would need more moisture to feed on, and would give out more oxygen, than we find on Mars.

Are we likely to encounter Martians? It is a thrilling question, although now that we know more about Mars it is less thrilling than it was a few years ago.

In 1877, the Italian astronomer Schiaparelli studied Mars very intensively through a low-powered telescope, and announced that in addition to the large markings that looked like seas there were finer markings, which, writing in Italian, he described as "canali". He used this Italian word merely to indicate channels of water, like the Grand Canal and the other canals in Venice, and did not mean to suggest that there were canals in the English sense, either straight lanes of water or the work of intelligent beings. Yet, as his description of them was translated into English by the word "canals", people began to argue that if there were canals, there must be intelligent beings to make the canals, and have so argued ever since.

Of late, however, doubts have been expressed as to the very existence of these channels or canals. There seems to be little doubt that astronomers see two different kinds of markings on Mars, which may properly be described as "subjective" and "objective". When the human eye is straining to its utmost to see things by inadequate light, there is an unmistakable tendency for it to see imaginary straight lines connecting up dark patches. An astronomer of my acquaintance illustrated this by putting an illuminated picture of a planet at the end of his garden, and asking his friends to observe it through a small telescope. A number were convinced they saw distinct black lines, like the Martian canals, although in actual fact there were no such lines to see; the simple explanation was that on the feebly lighted picture detail could only be seen with an effort, and the effort resulted in the seeing of non-existent lines. Another astronomer rubbed the canals off a drawing of Mars, and asked a class of schoolboys to draw what they saw. The boys at the back of the room put numerous canals into their drawings, and these were rather like the canals which had previously been drawn in by astronomers (Plate XXIX, p. 144). As the lines which the boys saw were imaginary, it is reasonable to suppose that those which the astronomers had seen were also imaginary.

Astronomers who claim to see canals on Mars usually draw these as straight lines on their maps, yet it is obvious that whether they really were straight on Mars or not, they could not look straight in all positions of the planet; a canal which looked straight when Mars was in one position would look curved, as a result of the curvature of the surface of Mars, when the planet had rotated to a new position (see fig. 65, facing p. 145). This again seems to indicate that the canals are mainly subjective illusions. The same conclusion is suggested by the fact that canals similar to those of Mars have been seen on surfaces where it is improbable that canals could exist, as, for example, on Venus, which is now believed to be covered in with thick clouds, on Mercury where water would boil (fig. 65, facing p. 145), and on satellites of Jupiter where it would freeze.

The camera is usually supposed to provide the final test of reality, and, although photographs of Mars shew quite definite markings, these do not resemble the supposed systems of canals. Perhaps this is not conclusive evidence, because photography is, for technical reasons, unsuited to the recording of very fine markings, and it is quite possible, as the canal observers claim, that these are best seen by the eye.

Taking it all together, the general accumulation of evidence and the general opinion of astronomers are equally against the supposed canals having any real existence. This does not of course prove there is no life on Mars, but it removes the main cause which has led a great many people to think there may be.

Thus, if we decide to take our rocket to Mars, I do not think we need trouble much about the prospect of meeting Martians. We are more likely to find an uninhabited and inhospitable desert, which may not shew quite the same extremes of climate as the moon, but may be even worse in some ways, since what warmth there is will never last for more than a few hours at a time.

If we leave Mars and continue our journey outwards into space, we shall find that we have to travel a very long way to pass from Mars to the next planet, Jupiter. Our journey may not be devoid of incident, since it will take us through the shoals of minor planets or asteroids, which have already been mentioned. The largest of these, Ceres, has a diameter of only 480 miles, which is less than a quarter of the diameter of the moon, and the only known limit to the size of the smallest asteroids is that set by the power of the observer's telescope. Smaller than the smallest asteroid we can see, there must be thousands that we do not see from the earth because they are too small to be seen. We may be able to see a number of them from our rocket, as this traverses the long distance between Mars and Jupiter.

Many of the asteroids are turning round in space, a complete rotation frequently occupying from 8 to 10 hours, and a number vary in brightness as they rotate. The reason for this variation is probably that the asteroids in question are irregular in shape, so that, as they rotate, the amount of surface they expose to our view continually varies. The huge gravitational pull of a big object such as the earth results in the object becoming very nearly spherical in shape, but a small body is not affected in the same way, and many of the asteroids are so small that gravitation can have done but little to mould them to a spherical shape. On many of them the pull of gravitation is so feeble that a good cricketer would be in danger of bowling all his balls off into space, and the batsman of making every ball a lost ball—lost for ever and ever, as the ball would itself become a new planet circling round the sun. Needless to say these asteroids are far too small to retain atmospheres.

At last we find ourselves clear of this swarm of asteroids, and are approaching Jupiter. Even from afar we see that it is far from spherical in shape; it is about twenty times as much flattened as the earth, so that at last we have found a planet of which we can truly say that it is flattened like an orange (fig. 66, facing p. 145).

The planet could not have been flattened like this if it had been standing at rest, for then its huge gravitational pull would have made it almost perfectly spherical. Thus it is not surprising to find that it is in rapid rotation, a complete rotation occupying a few minutes less than 10 hours. The flattening is quite adequately explained as the result of this rapid rotation, a point on the equator of the planet moving round the axis at a speed of about 28,000 miles an hour—as against 1040 miles an hour for a point on the equator of the earth.

We found Mars cold enough, but we shall find Jupiter incomparably colder. It is at more than five times the earth's

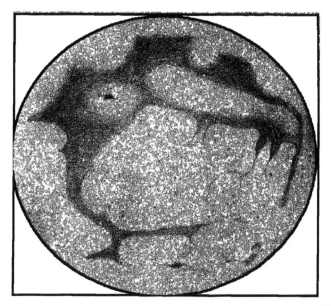

Royal Astronomical Society

Fig. 63. An attempt to examine the reality of the canals which some astronomers believe they can see on Mars The above drawing of Mars, which contains no canals, was put before a class of schoolboys who were asked to draw the picture as it appeared to them

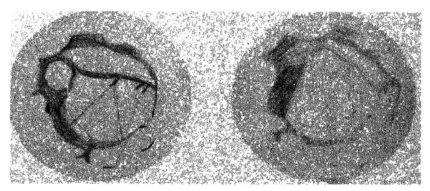

Royal Astronomical Society

Fig 64 Many of the boys put canal-like lines in their drawings, although no such lines appeared in the drawing from which they copied. Here are two of their drawings.

PLATE XXX

Fig. 65. The surface detail of Mercury after a
drawing by Schiaparelli

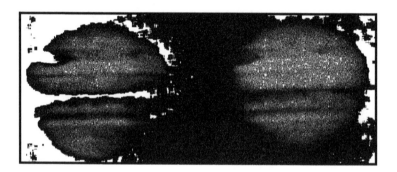

Mt Wilson Observatory

Fig. 66. Jupiter photographed in ultra-violet (left) and in blue light (right).

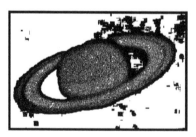

E. E. Barnard

Fig. 67 Saturn with its system of rings.

distance from the sun, so that 25 acres of its surface receive less of the sun's radiation than a single acre on earth. We can get a vague idea of the physical state of Jupiter by imagining the earth's supply of radiation suddenly reduced to a twenty-fifth or less. Its whole surface would very soon be frozen solid, and all activity would cease. We might naturally expect to find Jupiter in an equally dormant state, but it is not. Like Venus, it is completely covered in with clouds which are so dense that even infra-red light cannot penetrate them to any appreciable extent. These clouds shew remarkable and continuous changes. The best-known example is provided by the feature known as the great red spot. This was first noticed in 1878, and gradually increased until it attained a length of about 30,000 miles and a width of about 7000 miles—an area about equal to the total surface of the earth. It then became gradually more circular in shape and diminished in size, until it has almost disappeared by now. It is conceivable that this particular spot may have been produced by some special cataclysm, but other minor changes which are in progress all the time make it clear that Jupiter is no dead frozen mass. This is also shewn by the circumstance that belts of clouds in different latitudes rotate at different rates, those which are nearest to the equator rotating fastest.

All this activity was at one time regarded as evidence that Jupiter had a fairly high temperature, and so added substantially from some internal supplies of its own to the meagre supplies of heat it received from the very distant sun. We know now that this is not the case. Direct measurements shew that the temperature of Jupiter is at least 180 degrees below zero Fahrenheit, which is about what we should expect if it were warmed mainly by the sun's radiation, and had very little heat of its own.

With the temperature of Jupiter as low as this, it is clear that its clouds cannot be ordinary water vapour; they must consist of substances which remain in the vapour state at temperatures at which water vapour has long been frozen. As with the other planets, the composition of the atmosphere of Jupiter is studied by examining what wave-lengths are missing from sunlight which has passed into the atmosphere and out again. The observations are not easy to interpret, but they provide distinct evidence of two gases being present in the atmosphere of Jupiter, namely, ammonia and methane.

We all know ammonia as the stuff that draws tears to our eyes when we smell it, or when we unhappily break the bottle in which it is kept. Often also we can recognise its presence in smelling salts; it is the more efficacious but less agreeable ingredient, which the manufacturer tries to disguise by mixing with something more pleasant to the smell. We also find it useful to put on bee stings and mosquito bites, since, being very alkaline, it neutralises the acid of the sting, and so relieves our discomfort.

Methane is better known under its popular name of marsh gas. When vegetable matter decomposes under water this gas rises to the surface, where it may become luminous and appear as the "will o' the wisp" which is supposed to lure men on to destruction. It is also an ingredient of the fire-damp which is liable to explode in coal-mines, and also of the gases emitted at volcanic eruptions.

Neither of these gases is very attractive, and on the whole, the atmosphere of Jupiter appears to be what Hamlet would describe as "No other thing than a foul and pestilential congregation of vapours". We had better not take our rocket there; we should spend our time coughing, sneezing and crying. Moreover, as Jupiter contains 317 times as much substance as our earth, its gravitational pull is something to be treated with respect. We should not repeat the joyous experience we had on the moon of breaking our own and everyone else's athletic records without appreciable effort; on the contrary we should be very much concerned with the problem of how to support our own weight. The legs of a 12-stone man would have to stand as much strain as those of a 32-stone man on earth, and we might collapse under our own weight unless we did as the *Cetiosaurus* used to do on earth, and immersed ourselves in liquid to reduce the strain. If we are to travel round the universe without misadventure, we must not be above taking hints even from an extinct reptile (fig. 24, facing p. 44).

It is difficult to imagine that any planet could be less inviting than Jupiter, but Saturn appears able to fill the bill, and the planets still farther out—Uranus, Neptune and Pluto—are probably even less attractive than Saturn. We really know very little about any of these distant planets. Saturn appears to have rather less ammonia in its atmosphere than Jupiter, but makes up by having

Which is better, a clock that is right only once a year, or a clock that is right twice every day? "The latter," you reply, "unquestionably." Very good, now attend.

I have two clocks: one doesn't go *at all*, and the other loses a minute every day· which would you prefer? "The losing one," you answer, "without a doubt." Now observe the one which loses a minute a day has to lose 12 hours or 720 minutes, before it is right, whereas the other is evidently right as often as the time it points to comes round, which happens twice a day.

So you've contradicted yourself *once*.

"Ah, but," you say, "what's the use of its being right twice a day, if I can't tell when the time comes?"

Why, suppose the clock points to 8 o'clock, don't you see that the clock is right *at* eight o'clock? Consequently, when 8 o'clock comes round your clock is right.

"Yes, I see *that*," you reply.

Very good, then you've contradicted yourself *twice·* now get out of the difficulty as best you can, and don't contradict yourself again if you can help it.

You *might* go on to ask, "How am I to know when 8 o'clock does come? My clock will not tell." Be patient. You know that when 8 o'clock comes your clock is right; very good; then your rule is this· Keep your eye fixed on your clock, and the *very moment it is right* it will be 8 o'clock. "But—," you say. There, that'll do, the more you argue, the farther you get from the point, so it will be as well to stop. (Lewis Carroll.)

For further study, see:

J. Venn, *Empirical Logic*, Chap. X.
J. N. Keynes, *Formal Logic*, 4 ed. Appendix C, Chap. V, for a discussion of more complicated forms of agreement.

CHAPTER VI: GENERALIZED OR MATHEMATICAL LOGIC

1 State whether the relation in each of the following is transitive, intransitive, symmetrical, asymmetrical, one-one, one-many, or many-many.

a. He is the shortest man in the army.
b Joseph had the same parents as Benjamin.
c. Adam is the ancestor of all of us.
d. Impatience is not the characteristic of a good teacher.
e. Smith is a next-door neighbor of Jones.
f. Russia was defeated by Japan.
g. Romeo is the lover of Juliet.
h. The ticket agent is on speaking-terms with many notables.
i. Brown is an employee of Jackson.

2. Discuss the following·

"It is a profoundly erroneous truism, repeated by all copybooks and by eminent people when they are making speeches, that we should cultivate the habit of thinking of what we are doing. The precise opposite is the case. Civilization advances by extending the number of important operations which we can perform without thinking about them. Operations of thought are like cavalry

10. Show that an important condition for hypotheses is not fulfilled by the part of Freud's theory discussed in the following:

"[Freud declares that] 'the *libido is regularly and lawfully of a masculine nature, be it in the man or in the woman; and if we consider its object, this may be either the man or the woman.'* . . . Those individuals whose sex life seeks an object he calls the anaclitic type, and this is essentially a masculine type, since it is originally the woman who tends the infant. . . . *Later on he states that where woman is anaclitic or object-loving in her makeup, in that degree is she masculine.* This is a perfect example of the unassailable position, and has its analogs in much of male estimation of woman. Woman is primarily unintelligent, many men from Plato's time have said. But if they are shown a woman who is intelligent, their answer is, well, in that respect she is masculine!" [18]

For further study see:

A. D. Ritchie, *Scientific Method*, Chaps. III, IV, VI.
N. R. Campbell, *What is Science?*, Chaps. III, IV, V.
F. C. S. Schiller, "Hypothesis," in Chas. Singer's *Studies in the History and Methods of Science*, Vol. II.

CHAPTER XII: CLASSIFICATION AND DEFINITION

1. Examine the rôle played in modern astronomy by the classification of stars into constellations.

2. Attempts have been made to define several of the ethical concepts in terms of others taken as undefined. One such attempt consists in taking "better" as undefined. The following definitions are then offered:

A is worse than B	= B is better than A.	Df.
A is good	= A is better than the nonexistence of A.	Df.
A is bad	= A is worse than the nonexistence of A.	Df.
A is as good as B	= A is not better than B, and B is not better than A.	Df.
A is ethically indifferent	= A is not better than the nonexistence of A, and the nonexistence of A is not better than A.	Df.

Discuss these definitions from the point of view of (a) the psychological objective of definition, and (b) the logical objective.

3. What is the difference between natural and artificial classification?

4. Discuss the statement. "All description is classification."

5. What is the difference between a real and a nominal definition?

6. In what sense is it correct to say that the genus is part of the species, and in what sense that the species is part of the genus?

7. State the definition, a property, and an accident for each of the following: triangle, circle, star, animal, professor.

8. Point out the ambiguities in each of the following: bill, law, bolt, star, end, interest.

For further study:

J. Venn, *Empirical Logic*, Chaps. XI, XII, and XIII.

[18] Abraham Myerson, "Freud's Theory of Sex," in *Sex in Civilization*, ed. by V. F. Calverton and S. D. Schmalhausen, 1929, pp. 519, 520.

more marsh gas. It is even colder than Jupiter. On the other
hand, its gravitational pull is more like what we are accustomed
to, being only a sixth more than the earth's. It is the most
distinguished in appearance of all the planets, being surrounded
by a system of rings which look highly picturesque in a telescope,
and yet would bring disadvantages of their own if we took our
rocket to Saturn. For these rings consist of myriads of little
moons, each of which circles round Saturn in a very nearly
circular orbit (fig. 67, facing p. 145). Yet, as these little moons
are continually pulling on one another gravitationally, their
orbits cannot be perfectly circular, so that the tiny moons must
occasionally crash into one another. When this happens, broken
fragments of moon must be expected to fall onto Saturn itself,
with results which might be disastrous to a visiting rocket.

Before we leave these melancholy scenes, let us take a glance
at Pluto, the most recently-discovered, the remotest, and also
the chilliest, of all the planets. We know far less about it than
about any of the others, but it may perhaps be a sort of twin
brother to Mars, about equal in size and mass, but existing under
very different physical conditions. Each square yard of its
surface receives only a 1600th part of the radiation that a square
yard of the earth receives, so that its physical state is impossible
to imagine. Its gravitational pull is so feeble that it can hardly
have much of an atmosphere, but it may have more than Mars,
since its temperature is so far below that of Mars.

Surveying the whole scene, it seems likely that we may travel
through the whole solar system without meeting men like our-

selves, or even animals or vegetation of the kinds we know on earth. Yet on our own planet, the only part of the system with which we are familiar, life is so all-pervading that we can hardly believe that there are any physical conditions under which it is impossible for life to exist in some form or other. We find it in the coldest climates of our earth as well as in the hottest, in the depths of the sea, in the solid soil, and even in the streams of oil under the earth. In these various places it has assumed very different forms, each suited to its particular environment. This being so, we can hardly deny that it may have assumed still other forms on other planets, suited to the very different environments there. We have no right to say we shall find no life elsewhere than on earth, but it seems safe to suppose that if we do, it will be very different from any life we know—perhaps very different from any we can imagine.

The planets have proved so interesting that we have hardly left time to do more than glance at their satellites or moons. The earth has only its one solitary moon, but many of the planets are far richer—Jupiter, for instance, has ten, while Saturn has nine of respectable size, as well as the millions which form the rings. Uranus has four, Mars two, Neptune one, while Mercury and Venus, the two planets nearest the sun, have none, and the same is probably true of Pluto, the planet farthest from the sun.

Apart from the tiny moons in Saturn's rings, the nine planets have twenty-seven moons between them—an average of three moons apiece—so that the earth, with only one, may seem to have fewer than its fair share. This is true so long as we merely

judge by numbers. On the other hand, if we judge by weight,
our earth has more than its fair share; it has more moon substance
in proportion to its weight than any other planet.

We are all familiar with the tides which the moon raises in our
oceans. As the moon's distance is approximately thirty diameters
of the earth, its gravitational pull is a thirtieth part more at that
point of the earth's surface which is just under it than at the
centre of the earth. In the same way, it is a thirtieth part less at
the antipodes of this point. In fig. 68, we can represent the

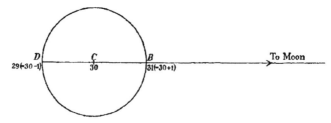

Fig. 68. Diagrammatic representation of the tidal force exerted by the
moon on the earth.

moon's pull at B, C, D, by the numbers 31, 30 and 29 respectively.
If we break up 31 into 30 +1, and 29 into 30 −1, we may suppose
that there is a uniform pull 30 all over the earth, with a pull +1
towards the moon at B, and a pull −1 towards the earth at D. This
latter is of course the same thing as a push +1 away from the
moon. The uniform pull 30 is exactly used up in keeping the
earth and moon in their proper orbits, so that we need not
trouble any more about it. On top of this uniform pull, however, are the pulls +1 and −1 acting at opposite sides of the earth.
These opposing pulls stretch the earth much as we might stretch

a piece of india-rubber by pulling in opposite directions with our two hands, and in this way cause tides. We have already seen that the earth is more rigid than steel, so that it yields less to the pull than the fluid ocean above. As a consequence, we are hardly conscious of any tides except those in this fluid ocean, yet the tides we see are really the difference between the tides in the ocean and those in the solid earth, the latter being quite small in comparison with the former.

If the little moon can stretch the big earth in this way, it stands to reason that the big earth must stretch the moon even more, and the same must be true of all the planets and their moons. We can never see our moon being stretched, because we can never see it broadside on, but we can see the process very clearly in the case of one of Jupiter's moons. A telescope shews that the moon which is nearest to the giant planet is so stretched out that it looks more like an egg than our idea of what a moon should be. In course of time this little moon will move in even closer to Jupiter. The nearer it goes, the greater Jupiter's pull will be, and the more stretched and egg-shaped the little moon will become—Jupiter is stretching it out more and more just as though it were a piece of rubber or elastic.

Yet we know that no piece of elastic can stand being stretched out indefinitely. It must snap in time—and so must the little moon. Calculation shews that the moon will at first snap into two distinct pieces, and Jupiter will have one more moon than now. But as these two new little moons will still be just about as near to Jupiter as the old moon was, they too will be strained and egg-shaped. In time they too must break up, and, as the process continually repeats itself, the number of Jupiter's moons will increase indefinitely.

We can say that Jupiter is surrounded by a sort of danger zone. When a moon or other body approaches this danger zone, it becomes egg-shaped; when it finally enters the danger zone, it is broken up—and if it stays within the danger zone for long enough it will be broken up into a vast number of tiny moons.

This is not mere guess-work, but the result of precise mathematical calculation. As soon as we know the gravitational pull either of a planet or of any other object, we can map out its danger zone. There are naturally different danger zones for different substances; a cloud of tenuous gas is in danger in regions where a rigid steely solid can venture in perfect safety. Now such calculations shew that the little egg-shaped moon of Jupiter is very near indeed to its danger zone. One of the little moons of Mars is also near—although not so near—to the danger zone of Mars, and one of Saturn's moons to the danger zone of Saturn.

This latter danger zone is of very special interest, because the millions of little moons which surround Saturn and form its rings are already inside it. It looks as if at some time in the past an ordinary moon had wandered inside the danger zone of Saturn and had been broken up into the millions of tiny moons which now form the rings. These rings are a standing proof that the danger zones have a real existence, and caution other bodies as to the fate awaiting them if they get caught by the gravitational pull of bigger masses. I have already reminded you of

Mr Kipling's story of how the elephant got his trunk. I have now tried to tell you the story of "How Saturn got its Rings"—perhaps not with Mr Kipling's grace and charm, but at least I believe that my story is a "really-truly" story, and not merely a "Just-So" story.

Our own earth, too, has its danger zone. So far the moon has kept well outside it, but in time the earth and moon must draw nearer together, and as they do so the moon will become more and more egg-shaped, until it finally crosses the danger line and begins to break up. It is only a matter of time until we have a fringe of rings like Saturn. In those far-off days we shall have lost our moon, but not our moonlight, for the myriads of tiny moons will still reflect the sun's light down to us at night; indeed, there will be even more moonlight than now, for a moon, like everything else, increases its total surface when it breaks up into fragments. There will also be moonlight all through the night. Still, life on earth will not be very comfortable, for every now and then two of the tiny moons will collide with one another, and their broken fragments will fall to earth like immense meteors, precisely as they must even now be falling on Saturn.

The solar system provides other evidence that these danger zones exist. We have already seen that comets do not move round the sun in circular paths like the planets, but in elongated oval curves which we call "ellipses". Usually a comet does not begin to be interesting until it has approached quite near to the sun. Then the sun's radiation beating down on it causes it to throw out a huge "tail", which may often be millions of miles long. The comet then becomes an interesting, beautiful, and even terrifying object.

Sometimes a comet will pass inside a danger zone, perhaps of the sun, or perhaps only of Jupiter or Saturn, and break up in consequence. Quite a number have been observed to break up into two pieces, while one has been observed to break into four. The most interesting story is that of Biela's comet, which broke in two while under observation in 1846. Six years later, when the comet's orbit again brought it near to the sun, the two pieces were observed to be $1\frac{1}{2}$ million miles apart. Since then, neither of them has been seen in cometary form, but the place where they ought to be is occupied by a swarm of millions of meteors, known as the Andromedid meteors. Occasionally these meet the earth in its orbit, and make a grand meteoric display—usually on or about November 27. It seems quite clear that since the comet first broke into two, both its pieces must have again traversed some other danger zone, and have been broken into innumerable small pieces in consequence. There are many other instances of comets disappearing as such, and being replaced by swarms of meteors.

Not only the sun, but of course every other star as well, exerts a gravitational pull, and so has a danger zone surrounding it. As the stars move onwards through space, it must occasionally happen that one wanders into the danger zone of another and more massive star. Events like those we have just been considering must then occur, but on a far grander scale. Just as the crocodile pulled the trunk out of the baby elephant, so the bigger star will pull a sort of trunk out of the smaller—a long nose or filament of gas, which will gradually break up into little pieces. It seems likely that sometime in the past the sun met with a misadventure of this kind, and that the pieces are our planets. So we can add another incident to our story—"How the Sun got its Planets".

The planets may also have met with misadventures of the same kind, wandering into the danger zone of the sun, and themselves getting broken up in turn. If so, we can write a further chapter—"How the Planets got their Moons". The saddest part of this chapter will be the story of one particular planet which seems to have met with a specially hard fate. It moved originally, we think, between Mars and Jupiter, but its motion took it into some danger zone, probably that of Jupiter. It began by breaking up, as though to make a few moons for itself, and ended up by being nothing but moons. At any rate nothing seems to have been left of it but tiny fragments, which are, as we believe, the asteroids or minor planets that we have already described. That one, Eros, which comes nearest the earth, is found to be shaped like an egg or a pear or a dumb-bell; perhaps it was about to break up still further, and just got away in time.

CHAPTER VI

THE SUN

So far we have been concerned only with the smaller of the objects in space. Smallest of all were the pellets of matter which we describe as shooting-stars when they fall into the earth's atmosphere; these are so small that we could hold thousands of them in each hand. The largest object we have discussed so far has been the giant planet Jupiter, with about eleven times the diameter of the earth. A box big enough to hold Jupiter would hold $11 \times 11 \times 11$ or 1331 earths—eleven each way. Yet even Jupiter is quite small in comparison with the sun, which we shall examine in the present chapter, and the sun is smaller still in comparison with the larger stars and other objects that we shall examine subsequently. Broadly speaking, the sun is as much bigger than Jupiter as Jupiter is bigger than the earth—Jupiter could contain more than a thousand earths, but the sun could contain more than a thousand Jupiters. To carry on the sequence, each of the blue stars we shall consider later could contain more than a thousand suns, while each of the "giant red" stars could contain more than a thousand blue stars. And each of certain nebulae which we shall discuss in our last chapter of all not only could contain, but actually does contain, thousands of millions of stars.

We can put this sequence in tabular form as follows, all the numbers of course being only very rough approximations:

PLATE XXXI

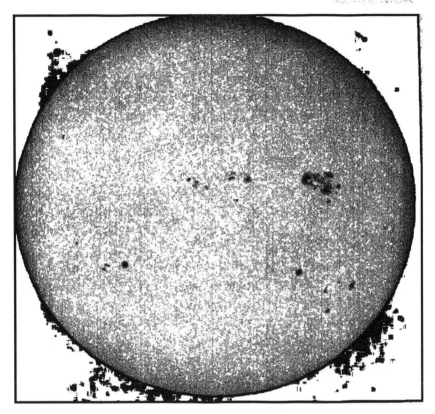

Greenwich Observatory

Fig 69   The sun photographed on August 12, 1917, when it exhibited very complicated and numerous sunspots whose total area was the greatest observed at any
time since 1870

PLATE XXXII

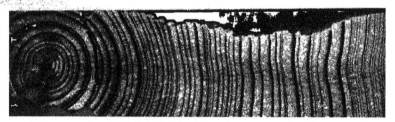

A. E. Douglass

Fig. 70. The cross-section of a beam of Douglas fir, shewing the variations of climate from A D 1073, when the tree started to grow, until A D 1260, when it was felled. Subsequently to A.D. 1260, the log was used as a beam in a dwelling, which afterwards fell into ruins. The beam was unearthed and studied in 1933, and is of value as filling in the weather record of the two centuries of its growth.

A. E. Douglass

Fig 71. The cross-section of a Scotch Pine felled at Eberswalde, Germany. The rings indicated by black spots were the growths of the years in which sunspots were most frequent from 1830 to 1906

Earth	1
Jupiter	1,000
Sun	1,000,000
Blue stars	1,000,000,000
Red stars	1,000,000,000,000
Nebulae	1,000,000,000,000,000

Let us imagine we take our rocket up once again, and examine the sun's surface from close quarters. Fig. 69 (p. 156) shews it as it might appear when we were well on our way. Perhaps the most noticeable feature is the darkening at the edge, or "limb" as the astronomers call it; at a casual glance it looks as though the edge of the sun were far less bright than the central parts of its surface. We can see the same darkening, even more clearly, on Plate XXXIV (facing p. 161). Now if the sun were solid or liquid, its surface would appear equally bright all over, as of course does the surface of a plain luminous globe. The apparent darkening of the limb provides a proof that the surface of the sun is gaseous.

We can see no other detail in our picture except groups of sunspots. These are rather unusual both in size and complexity; there are half a dozen at least which could swallow the earth quite easily, since, on the scale of the photograph, the earth would be only a grain of sand a twenty-fifth of an inch in diameter. Yet even these immense spots are nothing phenomenal in size, and occasionally sunspots appear which are large enough to swallow all the planets at one gulp.

We cannot see such sunspots as these every day, or even every year, but we can quite often see some spots. They do not come in a steady stream, but rather in gusts or waves, their numbers fluctuating up and down every 11 years or so. Sunspots were especially numerous in 1906, 1917 and 1928, and will be so again in 1939.

When we search the face of the sun for sunspots, we must be careful to look through dark glass, or at least through a piece of heavily smoked glass, or else we may find our eyes damaged beyond repair. Galileo, who was the first to study the spots on the sun, became blind in his old age, and attributed his misfortune to his gazing with unprotected eyes at the brightness of the sun.

People often discuss whether astronomical events, such as the coming of new or full moon, have any effect on the weather. Generally speaking, scientists are not able to trace any connection between the weather and any astronomical phenomena whatever, with the single exception of sunspots. There is, however, some evidence that the weather passes through a regular cycle having the same 11-year period as the frequency of sunspots. With the waxing and waning of the number of sunspots, the summers gradually change from being hot and dry to being cold and wet and then back again, the complete cycle taking about 11 years. Two instances will illustrate the nature of the evidence.

When a tree is cut down, we see a succession of concentric rings in the cross-section of its trunk, and each ring is known to be the growth of a single summer) We can tell how many years old the tree is by counting these rings) Yet, although the years must all have been of equal length, the rings are not of equal thickness. Some were formed in moist summers, when the tree grew luxuriantly and added profusely to its girth, others in dry summers which added but little to the size of the tree. By identifying the various rings with the successive years of the life of the tree, Professor Douglass claims that he can discover whether any particular year was dry or wet; the tree is, so to speak, a standing record of the weather it experienced throughout its life. Fig. 70 (facing p. 157) shews an interesting example. Now a careful study of

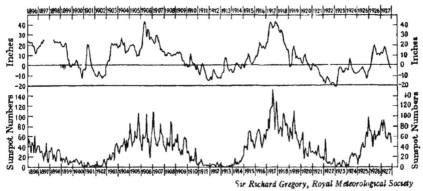

Sir Richard Gregory, Royal Meteorological Society

Fig. 72. The upper curve shews the height of water in Victoria Nyanza, while the lower shews the frequency of sunspots at the same time. We see that the curves keep almost perfectly in step with one another, demonstrating that sunspots have an influence on terrestrial weather.

such cross-sections of trees frequently shews that the rings change gradually in thickness in a cycle of 11 years, which coincides exactly with the sunspot period (see fig. 71, facing p. 157). The thickest rings were formed in those years when sunspots were most plentiful, and we see at once that abundance of sunspots goes with abundance of tree growth and so with moist summers.

Fig. 72 contains an alternative proof of the same thing. The lower curve shews the numbers of sunspots in the different years from 1896 to 1927, each up-and-down wave of this curve of course representing a single 11-year cycle of sunspots. The curve above this represents the height of water in Victoria Nyanza, the big fresh-water lake in equatorial Africa. We notice at once that the height of water in the lake keeps in almost perfect step with the frequency of sunspots, and so exhibits an 11-year cycle, just as the sunspots do. The water is of course highest after a wet year, providing proof that the weather is wettest when sunspots are frequent, and *vice versa*.

Although the frequency of sunspots changes slowly and gradually, so that the complete cycle is a matter of years, individual sunspots seldom last more than a few days. Plate XXXIII shews how much a big sunspot may change even within a single day. Plate XXXIV shews the gradual development of a very complicated group of spots, five of the six pictures having been taken on successive days. The spots move steadily to the right, not because they are moving across the face of the sun, but because the sun is rotating, and carrying them round with it. After the sixth day it is impossible to see the spots any longer, because the sun's rotation has taken them away from our sight.

An exceptionally big spot may occasionally disappear in this way, and subsequently come back, about a fortnight later, round the other edge of the sun. It was by measuring this motion of sunspots that Galileo first proved that the sun is rotating, and shewed that its time of rotation is about 26 days.

Passing over one of these spots in our rocket will be like passing over the funnel of a steamer in an aeroplane. We shall

PLATE XXXIII

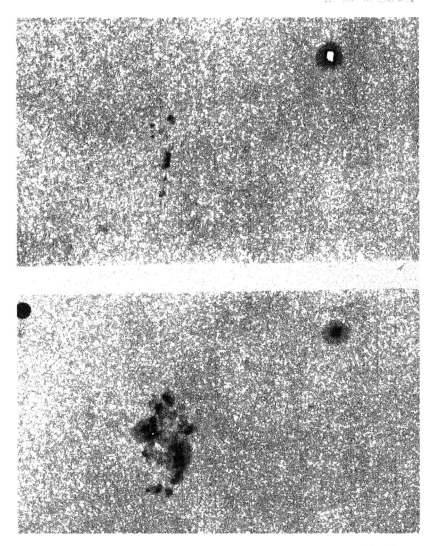

Fig 73. The development of a group of sunspots within an interval of 24 hours. The black circle on the lower plate represents the size of the earth

PLATE XXXIV

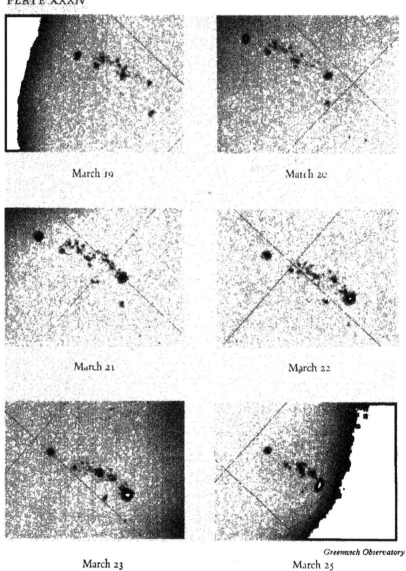

March 19

March 20

March 21

March 22

March 23

Greenwich Observatory

March 25

Fig. 74 A complicated group of sunspots, shewing motion, development and passage across the sun's disc in a period of 6 days (March 19-25, 1920).

notice a tremendous uprush of heated gas, and shall discover that
sunspots are of the nature of vent holes from which masses of hot
gas are shot out at terrific speeds. The fierce heat of the sun's
interior keeps the sun's outer layers in a state of continual agita-
tion; they may be compared to water which is made to boil
furiously by a hot fire underneath. We are all familiar with the
large bubbles of air and steam which force their way upwards
through boiling water. When they finally reach the surface, the
pressure which has so far compressed them is released, and they
expand and mix with the outer air. The material which comes
up in sunspots behaves in a similar way; as soon as it reaches the
sun's surface, the pressure on it is lessened, and it expands. As
a consequence of this expansion it becomes cooler for the reason
already explained (p. 60).

It is because the sunspots consist of cooler matter than the rest
of the sun's surface that they look black. Actually they are of a
blinding brightness, and look black only by contrast—because
they are less vivid than the hotter gases which surround them.
The matter which they eject is probably a mixture of complete
atoms and fragments of atoms, which may include electrified
particles of various kinds. These are shot out and travel in all
directions; after a day or two of journeying through space, some
of them will reach the earth, and, penetrating its atmosphere,
may produce a display of the Aurora Borealis. Later they may
ionise the air and so form the layers which reflect our wireless
waves back earthward and enable us to hear distant wireless
stations. We have already (p. 68) discussed what happens when

these electrified particles arrive on earth ; we are now seeing them at the beginning of their journey; we are present at the first of a long series of events, the last of which influences our lives on earth.

The column of gas which is ejected from a sunspot often rises to a great height above the surface of the sun, and is then described as a prominence. The matter which is hurled upwards from a big explosion or a volcanic eruption on earth may travel at a speed of hundreds of miles an hour, but the matter in these prominences is frequently hurled upwards at hundreds of thousands of miles an hour. Plate XXXV shews six successive photographs of such a prominence taken at intervals of only a few minutes. The last picture was taken within 2 hours of the first appearance of the prominence, and yet the ejected matter had already risen to a height of 567,000 miles above the surface of the sun; it must have travelled at an average speed of about 300,000 miles an hour.

Few prominences are as simple as the foregoing; usually their shapes are far more complicated, and change continually. The next series of pictures (Plate XXXVI) shews a prominence of a far more complex type, and its changes within the space of four successive days. As the sun turns round, we gradually discover that what at first appeared like a puff of smoke is an eruption of gas issuing from a sort of long crack in the sun's surface. This eruption was clearly a less explosive and altogether more leisurely affair than shewn in Plate XXXV.

The prominences are of very tenuous substance, being little

PLATE XXXV

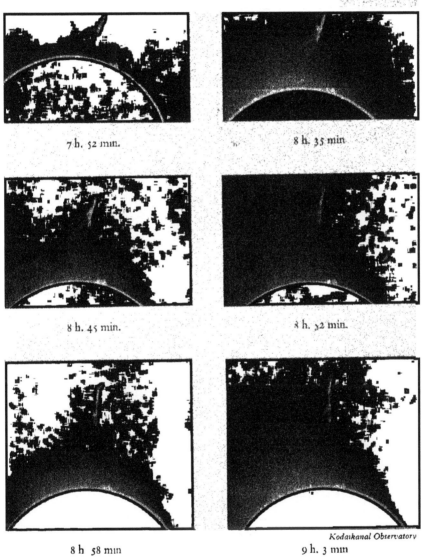

7 h. 52 min.

8 h. 35 min.

8 h. 45 min.

8 h. 52 min.

8 h. 58 min

9 h. 3 min

Kodaikanal Observatory

Fig. 75. A remarkable eruptive prominence seen on November 19, 1928. The pro—
minence attained a height of 567,000 miles in less than 2 hours.

PLATE XXXVI

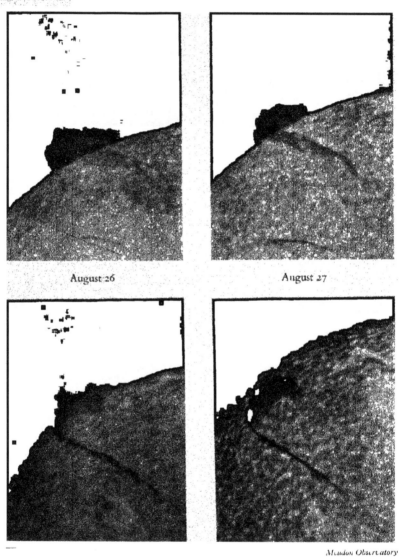

August 26

August 27

Meudon Observatory

August 28

August 29

Fig. 76. The development, and passage onto the sun's disc, of a solar prominence. The photographs were taken in calcium light (K3) on four successive days in 1929.

more than wisps of heated gas. They are also much cooler than the main body of the sun. For both of these reasons they look nothing like as bright as the proper surface of the sun, and so are usually lost from sight in the sun's glare, and cannot be seen under ordinary conditions. But when the moon passes in front of the sun and produces a total eclipse, the main body of the sun's light is completely shut off, the stars come out as at night, while the terrestrial landscape gets darker and darker, and finally assumes an ashy or slaty-purple appearance. Now is the time to see all the fainter lights surrounding the sun. The moment the last bit of sun is covered by the moon, the faint pearly light known as the corona flashes into view. The sun is surrounded, to a distance of hundreds of thousands of miles, by a tenuous atmosphere of molecules, atoms, and electrified particles, and the corona is simply this atmosphere seen by the light of the hidden sun. The light of the corona is less bright even than that of the prominences, so that prominences may frequently be seen shining through it. Figs. 77 and 78 (facing pp. 164 and 165) shew two photographs taken with different lengths of exposure at the eclipse of 1919.

Astronomers have devised means for seeing and studying all this, and much else, without waiting for an eclipse. We have seen how the surfaces and atmospheres of the planets can be studied in detail by the device of examining them in different colours of light—encouraging each colour of light to tell its own story. The surface of the sun can be treated by a similar method, not only with much greater ease, but also with far more profit. We no longer have to deal with a meagre quantity of reflected light, for the sun is itself pouring out a mixture of lights of all colours in such overwhelming profusion that it is easy to arrange for it to photograph itself in any colour of light we select. We need only break up the sun's light in a spectroscope, and then allow just that light which is of the precise colour we want, and no other, to pass out of the spectroscope into our camera. Yet there are very essential differences, which we must now consider, between this and the method used for the study of the planets.

Light and sound both consist of waves, and for this reason are similar in many respects. All the great noises of nature, such as the sound of a waterfall, a forest fire, a storm at sea, consist of mixtures of sound waves of all possible lengths. Of a different quality altogether from these confused torrents of noise, are the simpler and gentler noises we describe as musical sounds—cow bells on the mountains, church bells, the notes of a piano or violin. The confused torrent of sound contains waves of all lengths, but the musical sound contains waves of only a few lengths; that is why we find them pleasing to our ears.

It is the same with light. Sunlight, like the sound of a fire or a waterfall, contains waves of all lengths mixed, but there are other kinds of light which contain waves of only a few lengths— like a musical chord. If a beam of such light is passed through a spectroscope, we do not get a band of all colours, as with sunlight. We find that most colours are entirely absent, so that the "spectrum", instead of being a continuous band of colour ranging from red to violet, consists only of a few thin bright lines of colour here and there—we call it a line spectrum.

PLATE XXXV

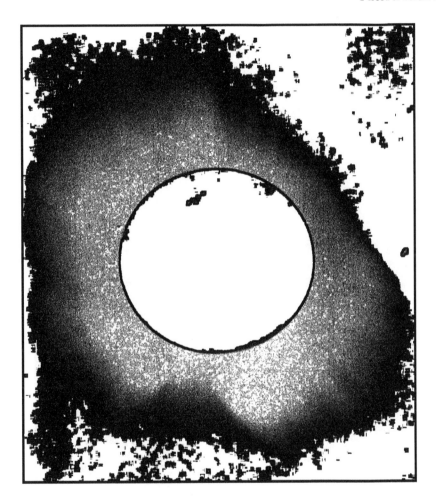

.1 C de la C Crommelin

Fig. 77 The solar corona, as photographed at the eclipse of May 29, 1919. A solar prominence can be dimly seen through the coronal light at the top left-hand edge (cf. fig 78, overleaf)

PLATE XXXVIII

A C de la C Crommelin

Fig. 78 The same as fig 77, but with a shorter exposure The prominence at the top left-hand edge is now seen quite clearly   It is found to have a length of more than 250,000 miles

Such spectra are usually emitted by the atoms of a chemically simple substance—what the chemists describe as an element. Not only so, but all the atoms of any one element, such as hydrogen, give out one chord of colours; while those of any other element, such as oxygen, give out another and quite different chord. Some substances give out light which is almost entirely of one single colour; naturally these are very popular for use in electric signs and luminous tubes.

Now suppose we put a small amount of any substance, say a pinch of ordinary table salt, into a hot flame, and watch what happens to the spectrum of the flame. A number of new lines will at once appear, which must of course have originated in the salt. Possibly we may be able to recognise some of the lines. Sodium, for instance, contains a very distinctive chord, consisting of two lines of excessively bright yellow lying side by side very close together. If we recognise these in the spectrum of our salt, we shall know that the salt contains sodium.

This method of tracing out the chemical structure of substances is described as "spectroscopic analysis", and provides an extremely sensitive test for the presence of many chemical elements. It will, for example, disclose the presence of a hundred-thousandth part of a milligram (1/3,000,000,000 oz.) of lithium. It is of course not necessary that we should put the chemical substance into the flame ourselves. If we can break up the light from any flame, no matter how distant it may be, into its constituent colours, we can tell something at least as to the composition of the flame; the light which travels from it to us brings with it a message as to what substances are producing the light. This makes it possible to investigate the composition of the sun and stars.

When Newton broke up sunlight into its constituent colours, he obtained a spectrum which he believed to be continuous, consisting of all conceivable colours arranged in order. But when Fraunhofer repeated the experiment in 1803, he was surprised to find that the spectrum was crossed by a number of dark lines, which he designated as *A, B, C, ..., K*. The spectrum was not continuous, but shewed brief gaps in the sequence of colours. There is a very simple explanation of these gaps.

Each atom in the sun's atmosphere is capable of emitting a chord of light consisting only of certain special and quite sharply-defined colours, but it cannot do this until it has first absorbed light of these same colours. [Generally speaking, the atoms of a hot substance are likely to be in what we describe as an "excited" state, in which they have stores of light of their own peculiar colours to emit. The atoms of a cool substance, on the other hand, are likely to be in an "unexcited" state, in which they are hungry for light of these colours.]

With this in our minds, let us fix our attention on the confused torrent of light which comes pouring up from the hot interior of the sun to the comparatively cool layers near its surface. It contains all colours of light, so that each atom in the comparatively cool atmosphere of the sun must be continually bathed in light of just those particular colours, among a host of others, which it is capable and desirous of absorbing. The atom

PLATE XXXIX

W. Huggins

Fig. 79. Spectrum of the sun.

W. Huggins

Fig 80 Spectrum of Vega.

W. Huggins

Fig 81. Spectrum of Sirius

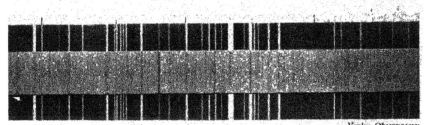

Yerkes Observatory

Fig. 82. Spectrum of ζ Ursae Majoris. The central band shews the spectrum of the star, the upper and lower bands being terrestrial spectra added for comparison so as to facilitate the identification of the lines.

Yerkes Observatory

Fig. 83. Spectrum of ζ Ursae Majoris. The central band shews the spectrum of the same star at a later date. Each line in the spectrum is now seen to be doubled shewing that the star is a binary system (see p. 187)

PLATE XL

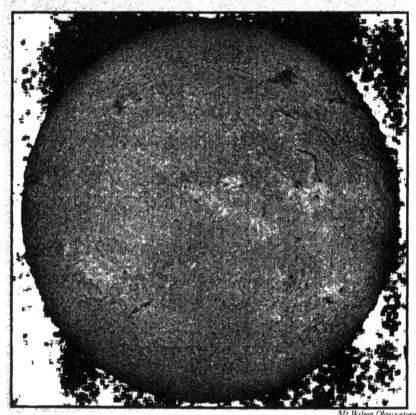

Fig. 84. The sun photographed in hydrogen light (Hα) This and the
photograph shown in Plate XXXI were taken simultaneously

naturally absorbs some of this light, and the main torrent goes on its way weakened in light of this particular colour. By the time it has run the gauntlet of all the hungry cool atoms which lie in wait for it in the sun's atmosphere, and finally emerges into space, it will be deficient in all the colours associated with these atoms—the colours they emit when hot, and absorb when cool.

For this reason, the spectrum of the sun is necessarily crossed by a number of dark lines and bands; these are not evidence of hot atoms emitting light in the sun's interior, but of cool atoms in the sun's atmosphere absorbing it. Fraunhofer knew of only eleven such lines, but the modern astronomer knows many thousands, and it is the same with other stars. On Plate XXXIX (facing p. 166), fig. 79 shews a fragment of the sun's spectrum, while figs. 80–83 shew spectra of other stars.

The position of these lines and bands provides the astronomer with an enormous storehouse of information, to which he returns again and again when he wants knowledge about the stars—how bright they are, how massive, how distant, how fast they are moving in space, how rapidly they are rotating, and so forth. The important point for us at the moment is that the colours of light which are missing in the spectra of the sun and stars can almost always be identified with the colours of light emitted by known substances on earth. When this can be done, we know that atoms of these same substances are at work in the sun's atmosphere, absorbing light as it comes out, and so preventing it from reaching us. It is precisely the method by which we discover that ozone is present in the upper atmosphere of the earth (p. 62).

(It is very significant that practically all the thousands of lines in the spectra of the sun and stars can be identified with the lines of substances which are known on earth.) This of course shews that the sun and stars are built up of the same kind of atoms as we are familiar with on earth—hydrogen, oxygen, nitrogen, iron, copper, gold, and so forth. If we travel to the sun or stars we shall expect to see many strange sights, but we must not expect to discover any new substances. The universe appears to be built of the same kinds of bricks throughout.

To return to our study of the sun's surface, suppose we now allow the sun to photograph itself in a certain colour of light which is emitted by, let us say, hydrogen atoms. We shall not, of course, get a photograph of the complete sun, nor even of all the hydrogen in the sun, but only of so much of the sun's hydrogen as is emitting this particular colour of light, and is at the same time near enough to the surface for its light to reach us. For instance, fig. 84 (facing p. 167) shews the sun photographed in a certain kind of hydrogen light, that of the line which Fraunhofer designated C, but which we now describe as $H\alpha$.

This photograph was taken at precisely the same moment as fig. 69 (facing p. 156). But in the earlier photograph all the colours of light were shouting together, and the only information which they tell us in common is "sunspots"; in fig. 84 the hydrogen light is quietly telling its own story, alone and undisturbed by the others. And an interesting story it is.

We learn how the hydrogen is not uniformly distributed in the sun, but occurs in a mottled formation of clouds, which look as though they were drifting, or being pushed about, rather like the clouds in the earth's atmosphere. Yet this comparison does not extend to size, since many of these clouds are far larger than the whole earth. Here and there, the mottled formation gives place to the long lines described as filaments, three of which can be seen near the top right-hand corner of the picture, while still another formation appears in the vicinity of sunspots. We notice that groups of sunspots affect an area of the sun's surface which is incomparably larger than the actual spots of blackness which superficial observation discloses. Fig. 85 shews the vicinity of a group of sunspots photographed in this same hydrogen light; we notice how closely the cloud formation is related to the positions of the dark spots.

Figs. 86 and 87 (facing p. 171) shew the sun photographed simultaneously in hydrogen light of a special kind (Hδ) and in calcium light (H$_2$). The pictures look very different, for the simple reason that one is a picture of hydrogen in the sun and the other a picture of calcium in the sun.

So long as the atoms of a gas are at rest and undisturbed they will not give out light at all. To make a gas light up we must do something to it, just as we must with an electric-light bulb, or a horse-shoe. For instance, we may pass an electric current through it—almost the only method available under terrestrial conditions—or we may heat it up; just as a solid horse-shoe begins to glow and then light up when we heat it, so it is with the atoms of a gas. It is in this way that the collection of atoms which forms the sun is made to give out light.

The blacksmith tells the temperature of a piece of hot iron by its colour. As he heats it up, its colour gradually changes—red hot, yellow hot, white hot, and so on. The same colour always denotes the same degree of heat, and this is true whether the emitting substance is iron or not.

It is much the same with a gas: we can tell its temperature from the kind of light it emits. All the atoms which have impressed their mark on the photographic plate reproduced in fig. 86 were excited in the same way, and so were all, within limits, at the same temperature. Thus their light, in recording itself on the photographic plate, has presented us with a picture of precisely those parts of the sun's atmosphere which are at this special temperature and no other, and so emit the special kind of light which we call $H\delta$. The atoms which are shewn in fig. 87 were at the different temperature at which the kind of light is emitted which we call H_2. We may say, then, that figs. 86 and 87 shew atoms of the sun which are at different temperatures.

As the sun's heat flows from a hot interior to a cooler surface, the hottest layers are naturally also the deepest. We have described our pictures as photographs of different parts of the sun which are at different temperatures, but we might equally have described them as photographs of different layers which are at different depths. When pictures of different layers of the sun are obtained in this way, the light they emit shews that many of their atoms are in a special state which is very well known to physicists—in

PLATE XLI

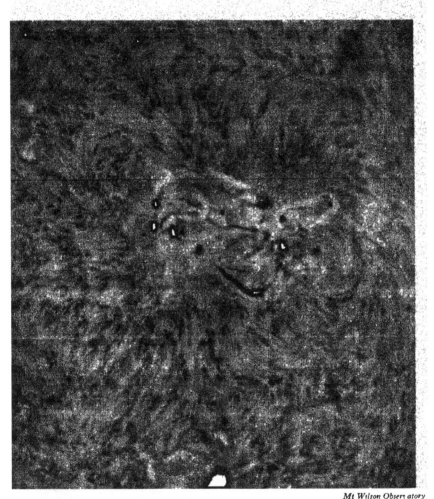

Mt Wilson Observatory

Fig 85   A complicated group of sunspots photographed in hydrogen light (Hα)

PLATE XLII

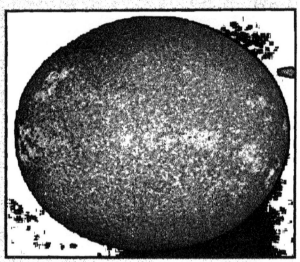

Mt Wilson Observatory

Fig. 86 The sun photographed in hydrogen light (Hδ)

Mt Wilson Observatory

Fig. 87. The sun photographed in calcium light (H₂) This and
the photograph above were taken simultaneously.

brief, they are partially broken up by the heat. And the deeper we go into the interior of the sun, the more the atoms are found to be broken up.

When we heat solid ice, it turns into liquid water; the molecules move more easily, because the bonds which have hitherto gripped them closely together have been broken down by the heat; as soon as they can slip quite freely past one another, the ice has turned completely into water. When we heat liquid water, it turns into gaseous steam; the bonds are still further weakened, so that the molecules can now move quite independently of one another. If we heat up the steam, even the bonds which hold the atoms together inside the molecules are loosened, so that the molecules themselves break up into atoms of oxygen and hydrogen. And if we could heat these atoms still further, until they reached the temperature of the sun's atmosphere, we should find that even the atoms themselves were breaking up —as they actually are in the outer layers of the sun.

If we take our rocket near to the sun's surface and analyse a sample of the sun's atmosphere, we shall find that it consists of atoms which are beginning to break up. But if we proceed inward to the sun's lower layers, we find the atoms breaking up more and more until, when we get near the centre, very little is left except completely broken atoms. It is a state of matter of which we have no experience, and we hardly know whether it is best described as solid, liquid, or gaseous.

We have seen how the pressure at the centre of our earth must be one of millions of atmospheres; that at the centre of the far more massive sun must be about 50,000 million atmospheres. Such a pressure as this packs the broken fragments of atoms so closely together that something like a pound of substance would go into a thimble. It is only because the atoms are broken up that they can be packed as closely as this.

It will never be possible to experiment with matter in this state in our laboratories—indeed, it would kill us if we tried. For, at a rough calculation, the temperature at the centre of the sun must be some 40 or 50 million degrees. And even a pinhead of matter at this temperature would radiate so much energy off into space, that we should need an engine of about 3000 million million horse-power to make good the wastage, and keep up the temperature of the pinhead of matter. This would emit its radiation in the form of a terrific blast against which nothing could stand. Quite close to the pinhead, the flow of radiation would produce a pressure of millions of tons to the square inch. This pressure plays its part in keeping the sun from collapsing, and plays an even more important part in the more massive of the stars, which it blows out until they are as tenuous as immense bubbles. Even a hundred yards away from our pinhead, the blast of radiation would be so strong as to blow over any fortifications which have ever been built, and it would speedily shrivel up any man who ventured to within a thousand miles of the pinhead from which it issued.

CHAPTER VII

THE STARS

We all know now that our sun is a very ordinary star, but it took men a long time to discover this. Perhaps this is not surprising, for certainly it does not look much like an ordinary star to us. The reason is, of course, that it is enormously nearer than any of the other stars.

We have seen how the ancients imagined the earth to be the fixed centre of the universe, round which everything else moved. The stars merely formed a background of light, against which they could map out the motions of the sun, moon and planets. They thought of the stars as attached to the inside of a hollow sphere, which turned round over the earth much as a telescope dome turns round over the floor of a telescope, or "as one might turn a cap round on one's head". And although a few of the more philosophical of the Greeks gave reasons for thinking that the earth moved round the sun, they had no means of making their opinions or arguments known to a wide circle of people, so that these were forgotten as the world gradually became submerged in the intellectual darkness of the Middle Ages. Then, in 1543, a Polish monk, Copernicus, advanced views and arguments which were very similar to those which Aristarchus of Samos had propounded 1800 years earlier, although the extent to which he was indebted to his Greek predecessors is not clear.

In brief, Copernicus declared that the sun, and not the earth, formed the centre of the solar system, that the earth was merely a planet, and that it, like all the other planets, moved round the sun.

Against this 1800-year-old thesis the eminent Danish astronomer, Tycho Brahe, as well as many others, raised an objection which was itself nearly 1800 years old. Indeed, Archimedes had previously brought forward precisely the same objection against the similar opinions of Aristarchus of Samos. The objection was, in brief, that if the earth were really moving round the sun in space, the apparent arrangement of the stars ought continually to change. If I walk about in a garden, I see the arrangement of the trees continually changing; one seems to move behind another, a third to step out into view, and so on. Yet a greenfly crawling about on a rosebud is not likely to notice any such changes in the arrangement of the trees—his rosebud is too small. Those who opposed Copernicus argued that, as no such changes were observed to occur in the arrangement of the stars, the earth must be standing still in space. They did not know that, viewed as objects in the celestial garden, the earth's orbit and even the whole of the solar system are less than the smallest of rosebuds. As Aristarchus had said 1800 years before Copernicus, the whole of the earth's orbit round the sun stands in the same relation to the universe as the centre of a sphere to its surface.

Nevertheless, when the positions of the stars are measured with the help of a powerful telescope, their apparent arrangement is found to be continually changing. The changes are of two distinct kinds. As the sun is continually forging ahead through the stars, and dragging us with it, the stellar scenery changes in the way in which terrestrial scenery changes when we drive through a forest. But besides this, the motion of the earth round the sun produces change of another kind. The sky of July will look different from the sky of January, because between January and July we shall have moved 186 million miles round the sun to the opposite end of the earth's orbit. When January comes round again, things will be back as they were in the previous January, because the earth will have completed its orbit, and we shall have come back to our original position relative to the sun.

If we continue to think in terrestrial terms, this 186 million miles' motion of the earth seems a fantastically long journey; on the astronomical scale it is so minute that for a long time astronomers were unable to detect the small apparent rearrangement of stellar positions which it caused. Indeed this was not detected until 1838, and it then became possible to measure the distances of the stars.

Exact modern measurements shew that the nearest stars are almost exactly a million times as distant as the nearest planets. We have already seen how sparsely scattered the planets are in the solar system; it now appears that space is even more empty of stars than the solar system is of planets. Five fruits placed in the five continents of the earth—an apple in Europe, a pear in Asia, a cherry in America, and so on—will give us a scale model shewing relation between the sizes of the stars and their distances from one another. We readily understand why the stars can only be seen as points of light, and we can further see that, even if the stars were surrounded by planets as is our own sun, these planets would be much too faint and also much too near to the central sun to be seen as separate objects.

If we take six wasps and set them flying blindly about in a cage 1000 miles long, 1000 miles broad and 1000 miles high, we shall again have a model of the distance of the stars. We can also make it represent the speeds of their motions if we slow down our wasps until they move only at about a hundredth part of a snail's pace.

We may be sure that, as the wasps fly about their big cage at this speed, they will not bump into one another, or even pass near to one another, at very frequent intervals. Yet it is most probably only when stars do this that planets like our earth come into existence—by the process we have already described (p. 155). For this reason, the birth of planets must be a rare event, and also, since the universe has not existed for ever, planets themselves must be very rare. People used to think of each star as giving light to, and supporting life on, a retinue of planets, but it now looks as though planets are the rare exceptions; at the most favourable computation, it seems likely that only about one star in every hundred thousand can have a family of planets to take care of.

We have already seen how greatly the stars differ from one another in apparent brightness. There are two distinct causes for this—the stars are intrinsically of different brightnesses in themselves, and are also at different distances from us. A star may look bright because it is near to us, as in the conspicuous instance of our own sun, or because it is a very bright object in itself, or of course from a combination of these two reasons.

As soon as we know the distance of a star, we can say how much of its apparent faintness or brightness is attributable to distance, and how much to intrinsic faintness or brightness. This makes it possible to compare the intrinsic brightnesses—or luminosities, to use the technical word—of the different stars.

The procedure is as follows. According to a well-known law of physics, light falls off as the inverse square of the distance; to take a simple illustration, if I walk to double my present distance from a street light, it will look only a quarter as bright. In the same way, if we place the sun at a million times its present distance, it will look a million million times less bright than now. At its present distance the sun has a brightness of twelve million million of the units we introduced on p. 95, so that if it receded to a million times its present distance, its brightness would be reduced to twelve units; we should still be able to see it, but only as a rather faint star.

A great number of the stars in the sky shine with more than twelve units of brightness, and all except three of these—Sirius, α Centauri and Procyon—are known to be more than a million times as distant as the sun. All these stars, then, must be intrinsically brighter than the sun. Sirius, α Centauri and Procyon are also known to be intrinsically brighter than the sun, and the same is true of most of the stars we can see with our unaided eyes. Broadly speaking, all the stars which look bright in the sky are intrinsically brighter than the sun.

Sirius, which appears the brightest star in the whole sky, is at a distance of 51 million million miles, or about 550,000 times the distance of the sun. If the sun were placed where Sirius is, it would have a brightness of only 40 units, whereas Sirius has a brightness of 1080 units. Thus Sirius is a very luminous star—twenty-seven times as luminous as the sun. Its brilliant appearance results from a favourable combination of the two factors which make for brilliance. It is both very bright in itself, and also very near, only one of the 5000 stars we can see without a telescope being nearer to us than Sirius.

Many of the nearest stars are of such low intrinsic luminosity that, in spite of their nearness, they cannot be seen at all by our unaided eyesight, but need a quite powerful telescope. The nearest of all known stars, Proxima, with a brightness of only a sixtieth of a unit, is so faint that it was only discovered quite recently. Its intrinsic luminosity is so low that it only gives out about a 20,000th part as much light as the sun, and even less heat. If it were put in place of our sun, the earth would become far colder than Pluto now is, and we should all be frozen solid in a very short time.

At the other end of the scale we find innumerable stars which are intrinsically more luminous even than Sirius, but look less bright in the sky because they are more distant. The brightest of all (S Doradus) gives out at least 300,000 times as much radiation as the sun, so that if it were suddenly to replace the sun, we should all be roasted in a fraction of a minute, and turned into vapour—sea, rocks, earth, and all—in a very few hours.

Nevertheless, the majority of stars prove to be fainter than the sun. Of the thirty stars which are nearest to it in space, only three are more luminous than the sun, while most of the remaining twenty-seven are very much less luminous. Even this is not the whole story, for it must be added that we happen to inhabit a part of the sky in which the stars are very distinctly above the average in luminosity.

We have seen how the apparent brightness of a star depends on two factors, namely, its nearness and its intrinsic luminosity. The latter of these two factors, the intrinsic luminosity of the star, itself depends also on two factors—the size of the star and the amount of radiation it emits from each square inch of its surface. We have found, for instance, that Sirius is twenty-seven times as luminous as the sun. But this leaves it an open question whether Sirius has twenty-seven times as much surface as the sun, or whether it is of the same size as the sun and gives out twenty-seven times as much radiation per square inch, or what other combinations of size and emission of radiation result in its total output being what it is.

The star's spectrum provides the means of answering this and all similar questions. For it tells us how much radiation the star emits from each square inch of its surface, and from this we can deduce the actual size of the star.

We have already seen that the quality of a star's spectrum depends on the temperature of the surface of the star. Different varieties of spectra correspond to different temperatures, with the result that, except for minute differences in detail, all spectra can be arranged in one single continuous series. As we pass from one end of this series to another, we are passing through a continuous range of temperatures of stellar surfaces. If we could gradually heat up the surface of a single star, we should find its spectrum passing through the whole of the sequence in succession. Indeed, sometimes nature performs this experiment for us; certain stars known as "variables" change in this manner spontaneously and of themselves, and we have only to watch the event happening to see the continuous sequence of spectra demonstrated in nature's own laboratory.

The amount of radiation which any surface emits also depends on the temperature of the surface; as a substance is heated up, it radiates out more and more energy. A really hot coal fire, such as we see in the firebox of a locomotive, may perhaps give about a quarter of a horse-power per square inch. The far hotter carbon in an electric arc may give as much as 6 horse-power per square inch.

When two stars shew identical, or very similar, spectra—as for instance Sirius and Vega (Plate XXXIX, facing p. 166)—we know that their surfaces must be at the same temperature and so are emitting the same amount of energy per square inch. Thus any difference in the intrinsic luminosities of two such stars can only result from a difference in their sizes. On the other hand, when two stars have different spectra, their surfaces must be at different temperatures and so must emit different amounts of energy per square inch. The spectra which form the sequence already mentioned may be identified with different temperatures and different emissions of energy to the square inch.

The spectra which form one end of this sequence indicate temperatures of not more than about 1400 degrees Centigrade, at which each square inch of the star's surface gives out only about a quarter of a horse-power—about the same as a really hot coal fire. We have seen how heating up a mass of iron causes its apparent colour to change in the sense of passing along the spectrum from the red end towards the violet. It is much the same with the stars, and these coolest stars of all are at such low temperatures that their radiation is almost entirely at the red end of the spectrum; they are in fact merely red hot. Many of them look red, or at least reddish, to the eye, so that they are frequently described as red stars.

About half-way along the sequence we come to spectra like that of the sun. These indicate a temperature of about 5600 degrees Centigrade, and at such a temperature each square inch of surface gives out about 50 horse-power. We can check the accuracy of this estimate in the following way.

If we measure how much sunshine falls on a square inch of the earth's surface, we can calculate first how much falls on the whole earth, and then how much is given out by the whole sun. If we divide this last number by the number of square inches on the whole surface of the sun, we can find how much sunshine is given out by each square inch of the sun's surface. We find that the energy equivalent of the sunshine given out by each square inch of the sun's surface is just about 50 horse-power—enough to run a powerful car all day and all night for millions of years, although of course not for all eternity, since even the sun's colossal stores of energy must come to an end some time. The area on which a single locomotive could stand would give out enough energy to run all the railways in the British Isles.

The spectra which lie at the remote end of the sequence indicate temperatures of perhaps 60,000 or 70,000 degrees Centigrade, so that each square inch of the star's surface will give out anything from 500,000 to 1,000,000 horse-power of energy—the amount of star that we could cover with a postage-stamp radiates out enough energy to run all the liners on the Atlantic Ocean. The main bulk of the radiation from these stars is invisible, lying far beyond the violet end of the spectrum. The visible radiation is largely concentrated in the violet end, so that the stars are often described as blue stars.

In this way the spectrum of a star informs us how much energy each square inch of its surface emits. Knowing the intrinsic luminosity of the star is of course the same thing as knowing the total amount of energy that its whole surface emits. A simple division will now tell us how many square inches of surface it has, from which it is an easy matter to calculate the diameter and size of the star.

The results of such calculations are very interesting, the more so as they shew that the sizes of the stars are not mere random quantities, but are closely connected with the physical states of the stars. Let us discuss this relation by working down gradually from the largest stars to the smallest.

The largest stars of all are without exception found to be red and cool. They only give out about a quarter horse-power of radiation per square inch, and so need a great many square inches to work off their heat. They are the immense stars, blown out like colossal bubbles by the pressure of radiation, to which we have already alluded (p. 172). We have recorded the disastrous consequences which would follow if *S* Doradus or Proxima were substituted for our sun. If one of these large red stars were to replace our sun, the results would be even more disastrous, since we should find ourselves inside it; the stars are larger than the whole of the earth's orbit. Indeed the largest yet known (Antares) has a diameter 450 times that of the sun—or about 400 million miles. We could pack about 60 million suns inside it, and there would still be room to spare. Our rocket, averaging more than 5000 miles an hour, took 2 days to reach the moon. If we tried to travel through the sun at the same rate it would take us a whole week, but if we tried to travel through this big star at the same rate, it would take 9 years. It is perhaps not surprising that astronomers describe these stars as "giants".

Let us now imagine that we measure all the stars, and place them in a row in order of size. We shall find that to a large extent we have also arranged them in order of colour. As we have just seen, the very large stars are all red; as we pass from these to rather smaller stars, we shall find the colour becoming less red. So it goes on, until finally we come to stars which are quite a lot smaller, having perhaps only ten or twenty times the diameter of the sun. These have only about a thousandth part as much surface as the red giants, so that to give out the same amount of radiation, they must give out a thousand times as much energy from each square inch. This being so, it is not surprising to find that these stars are at excessively high temperatures; they are the very hot blue stars of which we have already spoken.

We now seem to have used up the whole range of possible colours, and so of course of spectra, for stars, although we have only travelled a small way along the range of possible sizes—for the majority of stars are far smaller than the blue stars we have just described, with ten or twenty times the diameter of the sun. In actual fact as we pass to these still smaller stars, we find the range of colours and spectra merely repeating themselves. Instead of the smaller stars getting still hotter and bluer, we find them getting cooler and redder again, so that they not only have fewer square inches of surface, but also emit less radiation from each square inch. Clearly, they are far feebler stars than the red giants from which we started. In the end, we·come to stars which are just as red and cool as the huge giants, but far smaller in size. These are known as red "dwarfs"—with justice, since most of them are much smaller than our sun, and have only about a thousandth part of the diameter of the red giants. If we take a full stop on this page to represent one of these red dwarfs, the red giants will be represented by a cart-wheel.

So far we have found three main types of stars:
Very large (giants)—red and cool.
Middle size—blue and hot.
Very small (dwarfs)—red and cool again.

But the limit of smallness has not yet been reached, and stars even smaller than the red dwarfs are known to exist. The smallest of red dwarfs are still about the size of Jupiter or Saturn—only a thousandth part as big as the sun, but still a thousand times bigger than the earth. The smallest of all known stars are only about the size of the earth. These are described as "white dwarfs", because they are mostly white in colour, with spectra which usually correspond to temperatures of 10,000 degrees Centigrade or even more. Such high temperatures cause each inch of their surfaces to radiate intensely, and yet their surfaces are so small that their total radiation is very small indeed; they are so faint that only a few have so far been discovered.

We have already seen that the sun gives out light of all wave-lengths, although only about four octaves of light are given out in large amounts, and only one octave reaches us in abundance. We have also seen that a number of stars are far cooler than the sun. If we describe the sun as being white hot, these stars must be described as only red hot, and the radiation they give out is one or even two octaves lower than the radiation of the sun. If our sun had given out such radiation, we may perhaps suppose that our eyes would have adjusted themselves to it so that our visible spectrum would have been one or two octaves lower. We should not have been able to see our present green, blue, etc., at all, but only colours for which our languages contain no names, because we cannot see them. Grass, which now absorbs all colours except green, would look white, while the sky would look black. The scenery in general would look like the infra-red photograph shewn in fig. 37 (Plate XV, p. 76), while the infra-red pictures shewn in Plate XLIII suggest that many of the minor details of life would be different from what they now are.

These red stars, being cooler than the sun, give out light on the infra-red side of the sun's spectrum. Stars which are hotter than the sun naturally give spectra on the other, the ultra-violet, side. Sirius, for instance, with a temperature roughly double that of the sun, gives out a spectrum of light which is about an octave higher than the sun's. It may not appear to be so in an ordinary photograph, such as is shewn in Plate XXXIX (facing p. 166), because such photographs only shew the tail end running towards the red—most of the light is ultra-violet, and so is shut out by the ozone layer of our atmosphere. If Sirius had planets, the eyes of their inhabitants would probably have adjusted themselves to ultra-violet colours, for which again we have no names because we cannot see them. Life would be very different for such people. To take a trivial instance, glass is opaque to ultra-violet light, so that they could not use it for windows in their houses. On the other hand, it would do very well for the walls of the houses— except for the dangers referred to in the proverb. Air is nearly opaque to ultra-violet light because of scattering, and is quite opaque if it contains much ozone, so that if the Sirians had an atmosphere at all like ours, their sky would look perpetually black.

The hottest stars of all give spectra which lie three or three and a half octaves above that of the sun. If we want to find light of

PLATE XLIII

Ilford Co.

Fig 88   Portrait of a Hottentot, taken by infra-red and ordinary light respectively
The characteristic dark pigment is seen to be transparent to infra-red radiation

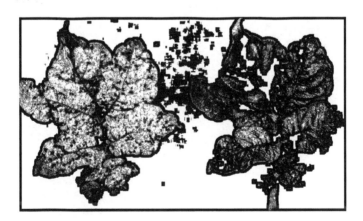

F C Bawden   Ilford Co

Fig 89   A potato leaf photographed in infra-red and ordinary light respectively.
The black marks in the infra-red picture indicate the presence of potato blight, and
cannot be seen in the picture taken by ordinary light.

PLATE XLIV .

Herbert Flower   Ilford Co.

Fig. 90. Star-fish as seen by X-radiation

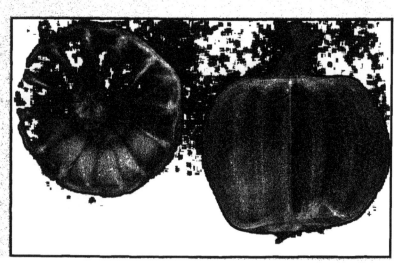

Herbert Flower   Ilford Co.

Fig. 91. Poppy-heads as seen by X-radiation

wave-lengths even shorter than this we must go inside the stars. If we sample the radiation a few thousand miles inside the sun, we shall find a spectrum like that of Sirius; a little farther in, the spectrum will have gone yet another octave up, and so on. At the centre of the sun, and probably of most stars, it is something like thirteen octaves up; the radiation here is of the kind we describe as X-radiation. Most substances are transparent to this, so that if we lived inside a star, shells, flowers, etc., would look like the photographs shewn in Plate XLIV, opposite.

So far we have been dealing only with those qualities of a star which can be seen by inspection, such as its temperature and size. We now pass to something more fundamental—the amount of substance the star contains, which we call its "mass". When we want to find out how much substance an object contains on earth, we usually weigh it, which means that we measure the gravitational pull between it and the earth. We can weigh the stars in much the same way, and so find out how much substance they contain.

Most stars pursue solitary paths through space, but occasionally we find them travelling in pairs, forming what is described as a binary system, or a double star. Each star grips the other fast in its gravitational pull, so that the two move through space together, each describing an orbit round the other. They keep together for just the same reason as the sun and earth; gravitation is too strong to permit of their separating—neither of them has speed enough to jump clear of the other.

We shall soon see that such binary systems are often very

interesting in themselves, but they are specially interesting as providing an opportunity for weighing the stars.

Each of the component stars of a binary system moves round the other somewhat as the earth moves round the sun, but with one very significant difference. The earth's mass is so much smaller than the sun's (1 to 332,000) that the sun's motion is hardly disturbed by the puny gravitational pull of the earth. In a true binary system, on the other hand, the two stars are much nearer to one another in mass. Consequently there is much more of an equal partnership in the matter of gravitational pulls, so that neither star is just making the other run round it, but rather the pair revolve about some point between the two. By noticing how much each star pulls on the other, we can find the ratio of their weights, and if we can also measure the dimensions of the orbit, we can find the actual weights of both the stars.

Sometimes the two stars which form a binary system are fairly similar in respect of size, colour and luminosity, so that the pair may properly be described as well matched. Such well-matched pairs are particularly frequent among the brightest and hottest stars of all. Indeed more often than not these brightest and hottest stars of all form constituents of binary systems. In such cases, the two constituents are often found to be very close indeed to one another; they may even touch, or—in extreme cases—overlap. It seems likely that stars whose constituents are as close as this originally formed a single mass, which has broken up as a result of spinning too fast for safety—rather as a fly-wheel is apt to break if it is spun too fast.

In other cases the two stars are exceedingly ill-matched and incongruous. A conspicuous instance is provided by Sirius, which forms a binary system with a white dwarf star. The principal star, the Sirius which shines so brightly in the sky, has a diameter half as large again as the sun, while its white dwarf companion has only a thirtieth of the sun's diameter. The red giant o Ceti provides an even more extreme instance of the same thing. It has about 400 times the diameter of the sun, and forms a binary system with a white dwarf companion whose diameter is unknown, but can hardly be more than a ten-thousandth part of that of the principal star. If the principal star is represented by a cart-wheel, its white dwarf companion is a mere grain of sand—perhaps only a speck of dust.

Even when there is an extreme disparity of size, the masses are often found to be fairly equal, the immense star perhaps having only five or ten times the mass of its minute companion. Generally speaking, it is likely that even white dwarf stars have masses which are comparable with ordinary stars. They resemble the earth in size, but the sun in mass. This of course means that the substance of a white dwarf star must be packed enormously more compactly than the substance of the sun. The average ton of matter in the sun occupies about a cubic yard, but the average ton of matter in an ordinary white dwarf would all go inside a thimble. By contrast the average ton of matter in the big star of o Ceti occupies about as much space as the interior of Waterloo Station.

Under such conditions as we encounter on earth, it is impossible to crush matter together as closely as it is in the white dwarf stars. The secret of these stars is that their atoms are broken up into their separate constituent particles. As we passed downwards in the sun, we saw the temperature becoming hotter and hotter, and the atoms more and more broken up (p. 171). At the centres of the white dwarfs, the temperature is incomparably hotter even than at the centre of the sun, so that the atoms are completely broken up, and can be packed into a very small space indeed.

The majority of binary systems do not belong to the sensational types which have so far been described, but consist of two components which are usually neither excessively close nor excessively dissimilar. For instance, fig. 92 (facing p. 196) shews photographs of the very ordinary binary star Kruger 60 taken in the years 1908, 1915, and 1920. When a large number of observations of the kind are available, it is easy to complete the orbit and then calculate the masses of the two constituents; in the case of Kruger 60 they are found to be one-quarter and one-fifth of that of the sun. Few binary systems are found in which the constituents have masses much smaller than these, but in the other direction we find masses ranging up to hundreds of times that of the sun.

The two components of the star Kruger 60 take fifty-five of our years to move round one another. Even this is a fairly rapid period of revolution for a binary star; many such systems have periods of thousands, and sometimes hundreds of thousands, of years.

At the other extreme are systems in which the period is very short, perhaps only a few days or even hours. Such systems cannot be seen or photographed as anything but single points of light, since the two components are too near to be seen as distinct stars in the telescope. Sometimes the orbits of such a binary system are so located in space that one component comes between the earth and the other component once in every complete revolution. At such moments the light of the second component is eclipsed and the total light of the star is temporarily diminished. A binary system of this kind is described as an eclipsing variable, and in favourable cases the observed changes in the total light may enable us to reconstruct the whole motion and calculate the size of the orbit, and also the diameters and masses of the two component stars.

We shall not of course see any eclipsing effect in a binary system unless the orbits of the components lie so that one component passes directly in front of the other as seen from the earth. But there are other ways of knowing that a system is binary.

When a train or motor-car rushes past us sounding its whistle or horn, we notice a fall in the pitch of the note as it passes by. This fall of pitch results from the wave-like nature of sound; our ears necessarily pick up more waves a second when the train is coming towards us than when it is receding from us.

Light also is of a wave-like nature, so that when a star is approaching us, our eyes pick up more waves per second than they would if the star were at rest, and the light of the star appears more blue in colour. If the star is receding from us, fewer waves are picked up, and the light appears more red in colour than it would normally do. Thus we can tell whether a star is receding from us or advancing towards us by studying its spectrum. When the spectrum contains sharp, clearly-defined lines, we can measure the amount by which they are displaced with great accuracy, and from this can deduce the exact speed of advance or recession of the star.

The spectral lines may be displaced by precisely the same amount year after year; in such a case we know that the star is moving towards or away from us at a perfectly uniform speed. In other cases the displacement varies continually, so that the star's speed of motion must continually be changing, and we conclude that the star is describing an orbit about a companion which is either entirely dark, or so faint that we cannot see its spectrum. Sometimes, as in the case of the star ζ Ursae Majoris whose spectrum is shewn on Plate XXXIX (facing p. 166), the spectra of both constituents can be seen, and we can then calculate the orbits of both as definitely as though we saw the stars themselves moving in space. Knowing the orbits, we can again calculate the masses of the constituent stars.

Thus we see that there are a great many ways of estimating stellar masses. Whichever method we use, the giant and blue stars are always found to be enormously more massive than the dwarf stars. The most massive star of which the weights are known with any certainty is a blue star known as Plaskett's star, in which the components each have about a hundred times the weight of the sun.

From methods such as those just described, we obtain a tremendous store of information as to the masses, sizes and temperatures of the stars. A few years ago astronomers could tell us very little about the stars except their names and positions in the sky, but they can now add a tremendous amount of information about each star. It adds enormously to the interest of our study of the sky if we can think of the stars in terms of their size, motion in space, weight, colour and other physical characteristics.

When we do this, we often find that a constellation is not a mere haphazard division of stars; its principal stars frequently prove to be very similar in their physical constitution, and at the same time are found to be all moving in the same direction with the same speed, thus shewing that they are physically connected.

A conspicuous instance is to be found in the stars of the constellation Orion. With the exception of the brightest star of the whole constellation, α Orionis or Betelgeux, practically all the brighter stars are moving in the same direction and at about the same rate. Also their physical characteristics are so similar that it becomes natural to compare them to a flock of well-matched birds. Apart from the exceptional star Betelgeux, the twelve next brightest stars are all exceptionally hot, exceptionally bright and exceptionally massive. They are all blue in colour, and belong to the class of stars whose members are exceptionally prone to break up into binary systems; and in fact all but one of these twelve stars are either known or suspected to be a binary. The brightest of the twelve, Rigel or β Orionis, is of special interest as being one of the most luminous stars known, its intrinsic luminosity being about 15,000 times that of the sun.

We obtain a slightly different story, although one in much the same style, when we turn from the constellation of Orion to that of the Great Bear. Again nearly all the stars are of one colour, but this time it is white. They form a group of stars which are less magnificent than those of Orion, a group which is much nearer to us and altogether a more homely affair, although still impressive enough. Six out of the seven of the stars which form the well-known Plough are white, and rather like Sirius in their physical characteristics. They are all larger, hotter, more luminous and more massive than the sun, although far less so than the Orion stars. Again, the brightest star in the whole constellation, a Ursae Majoris, or Dubhe, stands apart from the rest, being a rather large, cool, red star moving in a path of its own. Only three of the seven stars of the Plough are binary, for the stars of the Great Bear are already below the level of luminosity and temperature at which almost every star is binary.

We have already noticed that the conspicuous stars which form the constellations are mostly more luminous than the sun. They are also quite near home on the astronomical scale. For even the most luminous of all stars cannot be seen with our unaided eyes if they are at a great distance from us, and yet most of the conspicuous stars of the constellations can be seen very easily with our unaided eyes. Thus we may be sure that they are both exceptionally luminous and exceptionally near. Indeed it would be rather surprising if the most conspicuous stars in the whole sky did not owe their exceptional brilliance to a combination of favourable circumstances.

If we want to study the average star we must seek the aid of a telescope. We have already seen how this collects light, and so effectively increases the diameter of the pupil of our eye. A telescope with ten times the aperture of our eye ought to enable us to see every class of astronomical object to a distance in space ten times greater than that to which we can see without its aid. Thus, if the stars were uniformly distributed in space, we ought to see a thousand times as many stars as with our unaided eyes. With a telescope of twenty times the aperture of our eye, we ought to see eight thousand times as many stars, and so on indefinitely. When we actually perform this experiment, we find that the law holds good up to a certain distance only. After that it begins to fail; we see fewer stars than the law would lead us to expect, as though some stars were missing from their places. This must of course mean that space is not uniformly filled with stars. If we go far enough, we shall come to a limit where the stars begin to fall off, and we discover where this limit occurs by noticing where the law first begins to fail.

The five photographs of star-fields on Plates XLV–XLVII (following p. 196) will shew how the method is worked. The same field of stars is photographed with different exposures, these being adjusted so that each plate (except the last) shews stars three stellar magnitudes fainter than the plate before it. Now a star which is three stellar magnitudes fainter than another has approximately a sixteenth of the brightness of the latter (see

13-2

fig. 42, p. 97). It is easy to shew that if space were uniformly filled with stars each plate except the last would contain 64 times as many stars as the preceding; for the last plate the ratio would be 16.

Actually the increase in the number of stars on successive plates falls far below these numbers, shewing that the limit to the system of stars is touched quite early in the sequence.

The two Herschels, father and son, used a similar method for mapping out the shape and the limits of the system of stars to which our sun belongs. If the sun were at the centre of a globular mass of stars, the limit would obviously occur at the same distance in all directions. Actually the limit is found to occur at different distances in different directions.

If we are caught in a snow-storm when we are at sea, or crossing a flat plain on the earth, we may see snow-flakes surrounding us on all sides and apparently forming an opaque barrier of snow, while up above the sky may be comparatively clear. The reason for the difference is of course that the snow surrounds us for many miles in every horizontal direction, whereas it extends for one mile at most in the vertical direction.

The Herschels found that the stars were arranged like the snow-flakes in a snow-storm—i.e. in a flat disc—and concluded that the system of stars must be shaped somewhat like a snow-storm, or a coin or like a cart-wheel. They believed that the sun was somewhere near the centre, but we know now that they were mistaken in this, their telescopic power being utterly inadequate to reach anywhere near to the edges of the system.

PLATE **XLV**

E. E Barnard, Yerkes Observatory

July 1908   September 1915   **July 1920**

Fig 92 The binary system Kruger 60 (at top left-hand corner) as seen in the summers of 1908, 1915 and 1920 respectively. The two components perform a complete rotation round one another in 55 years, so that by 1963 they will be back in the position shewn in the figure on the left.

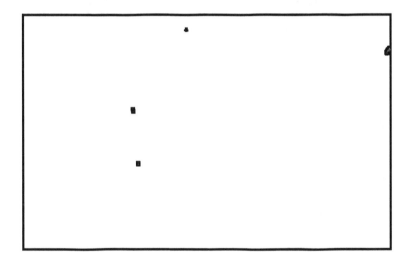

Mt Wilson Observatory

Fig 93. A small part of the constellation Auriga, shewing the only star brighter than the ninth magnitude (indicated by pointers). This forms the first picture of a sequence which is continued on Plates XLVI and XLVII, overleaf.

PLATE XLVI

Fig 94   The same field as in fig 93, shewing stars
down to the twelfth magnitude

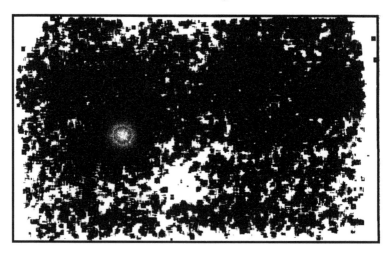

Fig. 95   The same field again, shewing stars down to the fifteenth magnitude   (The
pattern which is forming round the brightest star is of course merely an instrumental
defect.)

PLATE XLVII

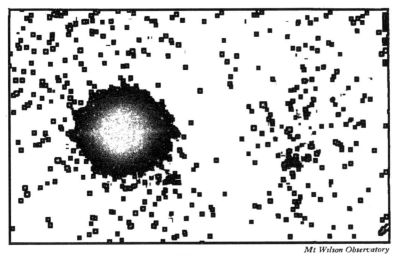

Mt Wilson Observatory

Fig. 96   The same field again, shewing stars down
to the eighteenth magnitude.

Mt Wilson Observatory

Fig. 97.   The same field again, shewing stars down
to the twentieth magnitude.

The sun is very far from the centre, although it is very nearly in the central plane of the system.

If we look in the direction which lies along the central plane of this coin or cart-wheel of stars, we are looking through the greatest possible thickness of the system and so ought to see an almost solid wall of stars, like the wall of snow-flakes we see when we look towards the horizon in a snow-storm. This solid wall of stars is the Milky Way, which we can see spanning the sky as a faint band of glimmering light on any clear, moonless night. The constitution of the Milky Way had been something of a mystery up to the time of Galileo, but his telescope at once shewed that it consisted of stars, as indeed Anaxagoras and Demo-critus had conjectured more than 2000 years earlier. These stars are so far away that we cannot hope to see them at all as separate individuals, but the light of millions of millions of distant faint stars combines to produce the illusion of a continuous cloud of light.

The sky we see without telescopic aid consists of this back-ground of very faint distant stars overlaid with a foreground of the few bright stars which form the constellations. Telescopic study at once connects up this background and foreground by shewing that a middle distance exists, consisting of stars which are both too faint to be seen individually and too sparsely scattered to form a continuous cloud of light. In this way the sun is found to be a member of a single system of stars which is shaped, as we have already said, like a disc or a coin or a cart-wheel.

We have already spoken of the surveyor's method of determining the distances of the stars—we travel over 186,000,000 miles from one end of the earth's orbit to the other and notice by how much the apparent direction of a star changes in consequence. Unfortunately, this method is only successful for quite near stars. The nearest star of all, Proxima Centauri, is at a distance of 25 million million miles; to avoid using big numbers we often specify this distance as $4\frac{1}{4}$ light-years, because light, which travels nearly 6 million million miles in a year, needs $4\frac{1}{4}$ years to travel from the star to us; we see the star, not as it is now, but as it was $4\frac{1}{4}$ years ago.

Now the surveyor's method enables us to find the distance of such stars as this with good accuracy, but it is, naturally enough, less successful with stars at a greater distance. It begins to fail badly for stars whose light takes more than a few hundred years to reach us, and is quite useless for stars which lie anywhere near the outer confines of our system of stars. Other methods must be found for determining the distances of these.

The most useful method depends on estimating the intrinsic brightness of a star from its general physical characteristics. For as soon as we know the intrinsic brightness of a star, a comparison with its apparent brightness will at once tell us the distance of the star.

There are three special types of star whose intrinsic brightness can be determined with fair, although varying, degrees of accuracy. We have already noticed that all the blue stars are very luminous, and in actual fact their intrinsic luminosity is found to depend almost exclusively on what we may call the degree or blueness, or in other words on the exact spectral type of the star. The same is true of the very large stars that we have described as red giants.

Consequently by studying the spectra of stars of either of these kinds, we can discover the intrinsic brightness of the stars and hence deduce their distances.

There is, however, a third class of stars whose distances can be fixed with even greater accuracy. These are the stars known as Cepheid and long-period variables; they do not shine with a steady light, but their brightness varies continually from day to day. This cycle of changes of brightness repeats itself at perfectly regular intervals, and the intrinsic luminosity of a star is found to depend almost exclusively on the length of the interval; stars which change most slowly are intrinsically the brightest; those which change most rapidly are the least bright. However distant such a variable star may be, we can measure the length of time from bright to bright or from faint to faint. This simple observation discloses the intrinsic brightness of the star, and from this we can deduce its distance.

Yet even with all these methods at our disposal it would be a difficult matter to map out the system of stars without adventitious aids. Such aids are provided by the objects known as "globular clusters". These are themselves minor systems of stars, far smaller than the main system, and yet each containing hundreds of thousands of stars. Each of these clusters contains great numbers of Cepheid variables, which make it easy to determine the distance of the cluster. When we know the distance of a cluster it is of course an easy matter to determine its size, and it is interesting to find that the globular clusters are almost exactly similar to one another in shape, size and general arrangement— we do not know why.

When the positions of these clusters are mapped out, it is found that they form a coin-shaped or disc-shaped aggregate; this is roughly circular in shape, and lies equally and symmetrically on the two sides of the Milky Way. It seems reasonable to suppose that the general arrangement and position of the system of globular clusters coincides with that of the system of stars, so that where the clusters come to an end, the system of stars also comes to an end. If so, the system of stars must have a diameter of roughly 200,000 light-years. But so far from its centre being near the sun, as the Herschels thought, it is something like 40,000 light-years' distant.

Thus we may think of the galactic system as a disc or coin or cart-wheel with the sun lying in its central plane, but perhaps a third of a radius out from the centre. The centre of the system is so remote that we cannot even see its brightest stars with our un-aided eyes; these can at the best only see stars whose light takes 3000 years to reach us. This explains why the bright constellation stars appear to be uniformly spread in all directions; we only see a tiny piece of the whole structure, and inside this tiny piece the stars really are spread fairly uniformly.

It has recently been discovered that the motions of the stars are neither random motions nor uniformly arranged; indeed it now seems to be established that the whole system is rotating round a centre, much as a cart-wheel rotates about its hub. This rotation of the great wheel of stars whirls the sun through space at a rate of about 200 miles a second, yet the wheel is so vast that the sun must travel at this speed for about 250 million years before it has made one complete circle round the hub.

Such a rate of spin as this is almost inconceivably slow—one turn in 250 million years. To try to realise what it means, let us compare our rotating cart-wheel to the hour hand of a clock which turns completely round once in 12 hours. If we now slow the hour hand down until it turns at the same rate as the system of stars, the jump which at present occurs every second would take more than 5000 years—almost the whole of human civilisation. Yet a study of the ages of the stars seems to shew that our wheel must have made thousands, and perhaps hundreds of thousands, of complete revolutions.

The sun would fly off the whirling wheel into space, like the speck of mud off a bicycle wheel, were it not that the gravitational pull of the other stars restrains it. This gravitational pull keeps the sun moving in an orbit, just as the gravitational pull of the sun does the earth. And just as our knowledge of the earth's orbit makes it possible to calculate the mass of the sun, so our knowledge of the sun's orbit makes it possible to calculate the total mass of the stars which constitute the great wheel. We find that the number of stars in the wheel is certainly greater than a hundred thousand million, and may well be double this.

Our unaided eyes can distinguish at most about 5000 of this multitude of stars as distinct points of light—one in 40,000,000. Thus for every star that we see as a star, there must be 39,999,999 others that are either completely invisible or are merged in the general faint glimmer of the Milky Way. There are about 2000 million inhabitants of the earth, so that if the stars were divided equally among all the earth's inhabitants, there would be about 100 for each person. Yet if we choose our stars by drawing lots at random, each of us will find that there are odds of about 400,000 to 1 that he will not be able to see a single one of his stars without using a telescope.

CHAPTER VIII

THE NEBULAE

The moon and planets look very conspicuous objects in the sky, but we know that these are very near neighbours which only look bright and big because they are near. For the rest our unaided eyes can see nothing of the universe except stars.

A small telescope or field-glass will shew us more stars in abundance, but it will shew us something else as well. A new class of object comes within our ken, the fuzzy indefinite patches of faint light which we describe as "nebulae".

The word "nebula" is of course the Latin word for a mist or cloud. In the early days of astronomy it was used indiscriminately to describe any object of misty or fuzzy appearance—any object, indeed, which did not exhibit a clear outline. Since then it has been found that the nebulae fall into three distinct classes.

The first consists of objects known as planetary nebulae, which lie entirely within our system of stars. It is now known that these are themselves stars which, for reasons not altogether understood, have become surrounded by very extensive atmospheres. Examples are shewn in fig. 98 (facing p. 204). We described the red giant stars as large, but when their atmospheres are counted in, these stars are beyond all comparison larger. Our rocket, travelling at 5000 miles an hour, would take 9 years to travel through the biggest of red giants, but about 90,000 years to travel through one of these planetary nebulae. This means that if we regard the planetary nebulae as stars, they are some 10,000 times larger than the largest stars we have yet mentioned.

Strictly speaking, these nebulae are the atmospheres of stars rather than the stars themselves. Peering through these atmospheres, we see the stars themselves at the centres of the nebulae, and these are, if anything, more remarkable than the vast atmospheres which surround them. To begin with they are surprisingly small, with an average diameter of only about a fifth of that of the sun. Their surfaces are at excessively high temperatures, which range up to about 70,000 or 75,000 degrees Centigrade. These are the highest temperatures that can actually be observed in the universe, although we know that the interiors of stars—which we cannot observe—must be at still higher temperatures. In a sense the temperatures we have just mentioned are themselves internal temperatures, because they are measured at the bottom of the big atmospheres surrounding the stars and not at their surfaces. These small sizes and excessively high temperatures shew that the central stars of the planetary nebulae belong to the same general category as the white dwarfs we have already discussed (p. 185).

The second class of nebulae also consists of objects which lie within the system of stars bounded by the Milky Way. The nebulae of the first are atmospheres surrounding single stars; those of the second class may be described as atmospheres surrounding whole groups, and sometimes even whole constellations, of stars. Fig. 99 shews the familiar stars of the Pleiades, photographed with a long exposure. A casual glance at these

PLATE XLVIII

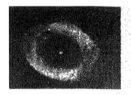  

Mt Wilson Observatory

Fig 98. Three planetary nebulae—N G C 6720 (the ring nebula
in Lyra), N.G.C. 2022, and N.G.C 1501.

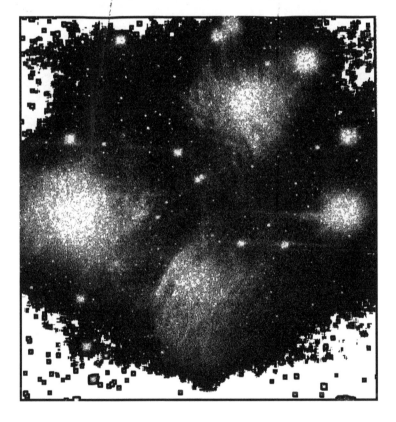

Karalyn, Kosmolgurer

Fig. 99. The stars of the Pleiades, with the nebulosity surrounding them.

PLATE XLIX

Fig. 100. The nebulosity which surrounds, and is lighted up by, a single
star in the constellation of Auriga.

stars, either with our unaided eyes or through a telescope, shews no nebulosity of any kind, but when the constellation is photographed with a long exposure each star is found to be surrounded by a nebulous cloud of light. Fig. 100 (opposite) shews the intricate detail of the nebulosity surrounding a single star.

With still longer exposures the nebulosities surrounding the different stars would join up to form a continuous cloud of light and we should find an enormous number of stars all immersed in one great unbroken sea of nebulosity. An example of such a sea of nebulosity is shewn in fig. 101 (facing p. 206). In many cases the nebulosity does not take the form of clouds of light, but of patches of darkness, a conspicuous example being shewn in fig. 102 (facing p. 207). It seems fairly certain that these dark patches are produced by absorbing matter which shuts out the light of the stars behind, and the absorption may well be of the same general kind as produces dark lines in stellar spectra and deprives our own atmosphere of its ultra-violet radiation. The light is absorbed by cool gas but is emitted by hot.

These nebulae look very sensational, but only in the way in which the moon and planets may look sensational—because they are comparatively near to us. The nebulae of the third class, to which we come next, are sensational in themselves. A planetary nebula may give out ten or perhaps a hundred times as much light as the sun, one of the "galactic" nebulae just described may give out perhaps hundreds or thousands of times as much, but the third class of nebulae, the "extra-galactic" nebulae, give out thousands of millions of times as much. They are enormously larger than the galactic nebulae in size, but look smaller and less impressive because of their great distance from us.

These three kinds of nebulae are so different in shape and general appearance that there is usually no difficulty in distinguishing between them. But their spectra provide a further means of discrimination, if this is needed. When the light of either the planetary or galactic nebulae is analysed in a spectroscope, it is found to give the same spectra as the various kinds of atoms we know on earth. This shews that these nebulae are mere clouds of luminous atoms—gas lighted up by the stars embedded in them.

The extra-galactic nebulae, on the other hand, give spectra like those of the stars. It is, then, natural to suspect that these are clouds, not of atoms but of stars. For a long time, this was nothing more than a plausible conjecture, but there can no longer be much doubt as to its truth. For, just as Galileo's telescope broke up the Milky Way into separate points of light which he at once identified as stars, so the modern high-power telescope resolves the outermost regions of these nebulae into separate points of light, which may, without hesitation, be identified as stars.

There can be no reasonable doubt that they really are stars, for they reproduce practically all the characteristics of the stars of our own system. Many, for instance, do not shine with a steady light, but fluctuate in the same characteristic and quite unmistakable way as the Cepheid variables of our own system. Quite recently other objects have been detected, similar to objects with which we are familiar in our own system of stars. Not only have variable stars of all kinds been found, but also "novae" or new stars

PLATE I

Fig. 101   Nebulosity in the constellation of Sobieski's shield.

PLATE LI

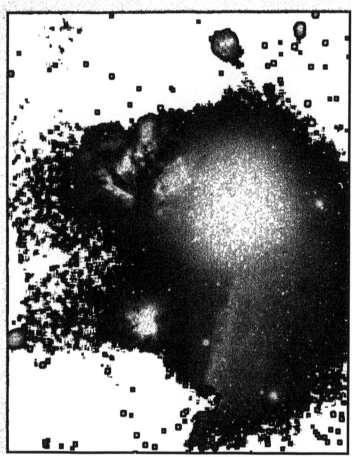

Kisilir, Forcalquier

fig. 102. Nebulosity in the constellation of Orion The bright object half-way up the plate is the star ζ Orionis, the southernmost of the three stars of Orion's belt The photograph was given an exposure of 11 hours, which is enough to shew all details of the clouds of obscuring nebular matter.

which suddenly flash out to thousands of times their ordinary brightness, and then, after undergoing several fluctuations of brightness and darkness, become faint again. Globular clusters have also been discovered very similar to those of our own galactic system. There is, then, no reason to doubt that these extra-galactic nebulae are, in part at least, systems of stars very similar to those of our own galactic system.

We have seen how variable stars and globular clusters occur throughout the galactic system of stars, and so make it possible to estimate the distances of the remotest parts of this system. The distances of the nearer nebulae can be estimated in precisely the same way. Cepheid and other variable stars can be recognised in these nebulae by the peculiar character of their light variation. They behave in just the same way as variable stars nearer home, but look enormously fainter because of their greater distance. And, as we have already seen, the difference in faintness at once tells us the difference in distance.

In this way, it has been estimated that the two nearest nebulae are both about 800,000 light-years' distant—the light by which we are now seeing them started on its journey through space about 800,000 years ago, when man was first appearing on earth. Fig. 104 (facing p. 208) shews one of these two near nebulae, the great nebula in Andromeda. In spite of its immense distance, it occupies quite a large part of the sky; if the full moon were in the same photograph it would only appear the size of a sixpence. And even this does not shew the whole size of the nebula. The more it is studied the larger it is found to be, and already it has been found to extend to several times the dimensions shewn in the photograph.

An object which is at so stupendous a distance and yet fills up so much of the sky must clearly be of immense size. Our rocket which took 2 days to reach the moon, would have taken a week to travel through the sun, 9 years through an ordinary big star, 90,000 years through a planetary nebula—but it is not much good saying how long it would take to travel through this nebula. Actually the nebula is about 100,000 light-years in diameter, so that the time would be about 12,000 million years. We should have to enlarge our photograph of the nebula to the size of all Europe before an object of the size of the sun could be seen in it.

We notice that this nebula has a sort of cart-wheel shape, such as we have already attributed to our own galactic system of stars. Indeed, the size, shape and general constitution of this nebula combine to suggest that it may be very similar to our own system. This, and a large number of other nebulae, are not only like cart-wheels in shape, but are also found to be rotating like cart-wheels about their hubs or centres—again like our own system of stars. Each wheel is held together as a compact structure by the gravitational attraction of its parts, so that we can calculate its mass by the method we have already used to calculate the masses of the sun and of the galactic system, although it is not possible to claim much accuracy for such a calculation. The Andromeda nebula is found to turn about the hub of its wheel once in every 20 million years, and from this it is

PLATE LII

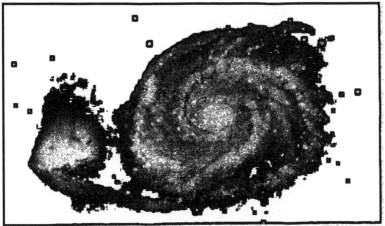

Mt Wilson Observatory

Fig. 103 The "whirlpool" nebula in Canes Venatici

Yerkes Observatory

Fig. 104 The great nebula in Andromeda.

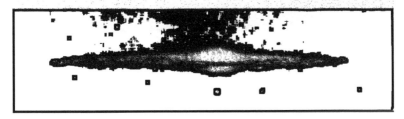

Mt Wilson Observatory

Fig. 105. A nebula (N.G.C. 4565) in Berenice's Hair.

Three nebulae, all of characteristic cart-wheel shape, and probably somewhat
similar in structure, viewed from different angles.

PLATE LIII

Mt Wilson Observatory

Fig 106 A close group of nebulae in the constellation of Pegasus. The nebulae near the centre of the plate all look of about the same size and brilliance, and so are all at about the same distance—they form a close group in space. Other nebulae, which look smaller and fainter, are probably at a greater distance, and so are not physically connected with the principal group.

calculated that its mass must be equal to that of several thousands of millions of suns—it is safest not to say precisely how many.

The extra-galactic nebulae are not all of cart-wheel shape; indeed they shew considerable diversity both of shape and general appearance. It is found however that almost all of them can be arranged in a single continuous sequence. This sequence begins with nebulae which are fuzzy in appearance, globular or nearly so in shape, and in which no stars can be discerned; it ends with pure clouds of stars like our own system. Only the nebulae in the latter half of this sequence are shaped like cart-wheels, and here the comparison is specially appropriate, since many of them are rotating around a sort of central boss or projection, which looks surprisingly like the hub of a cart-wheel. This cart-wheel shape may be more or less disguised when the nebulae happen to be seen from unsuitable angles (see Plate LII, facing p. 208), but it is very obvious when we view them edge-on, as in the nebula shewn in fig. 105. When we allow for actual nebulae being seen at all possible angles, we find that the sequence in question is simply an arrangement in order of flatness, the shapes ranging from spheres to cart-wheels.

When we take a walk through a forest of oaks, we come upon trees of all sizes, ranging from full-grown forest trees down to saplings, and even to young shoots just growing out of acorns, and acorns lying on the ground. Here again we find that all the stages we encounter can be arranged in order to form one continuous sequence, starting with the newly-fallen acorn, passing through the sprouting acorn, the baby oak tree, the sapling, and the young tree to the full-grown forest tree. We naturally suspect that the different appearances may represent different stages of growth, and so constitute an "evolutionary sequence". But this must remain only a suspicion; oaks grow slowly, and we cannot wait long enough to watch the change occur.

It is the same with the nebulae. Any appreciable change must occupy millions of years. Thus we cannot wait to watch them change, but we may conjecture that as they change they move on from one state to the next in the sequence. If this is so, the sequence becomes a sort of cinematograph film, exhibiting the life-history of a nebula. All the nebulae which come earlier than any particular nebula in the sequence are pictures of what this nebula has already been at some time in the past; those which come after it are pictures of what it will be at some time in the future.

The sequence of nebular shapes has a further interest, since calculation shews that it almost exactly coincides with the sequence of shapes which a huge ball of gas would assume as it gradually shrank, increasing its speed of rotation as it did so. The faster the ball of gas rotates, the flatter its shape—just as with the planets of the solar system. In time the shape becomes so flat that it cannot flatten any more; a further increase in the speed of rotation then causes matter to fly off from the equator—as we imagined might happen on our earth (p. 10) if this were set spinning fast enough. We imagine the rim and spokes of the cart-wheel to have been formed in this way, the hub being the much-flattened remains of the original ball of gas. The final end of this sequence is of special interest, since by the time it is reached the whole of the gas is condensed into detached globules, and calculation shews that each globule would have about the same mass as an actual star. It becomes natural to conjecture that each nebula started as a rotating mass of gas, that this mass passed through, or will pass through, the sequence of changes we have already described, and ended, or will end, as a cloud of stars. Thus the nebulae are the birth-places of the stars; in them rotating masses of gas are moulded into stars such as we know and find in our own galactic system.

If these conjectures are sound, we can trace our earth back to the sun, and the sun back to a nebula, but how did the nebulae themselves come into being?

Most cosmogonies have taken as their starting-point the supposition that the universe started as a chaotic mass of gas. It can be shewn that such a mass of gas could not stay uniformly spread throughout space. The cloud of steam from a kettle or from the chimney of a locomotive does not stay uniformly spread out, but tends to condense into tiny drops, and we find that it would be the same with gas of any kind spread through space. A uniformly spread gas, whatever its nature, would be unstable in the sense that any slight disturbance or irregularity would tend to increase indefinitely instead of smoothing itself out. Finally, the whole mass would condense, or break up, into detached masses of denser gas. Calculation shews that these would be on something like the scale of the actual nebulae, and would form at about the average distance apart of the observed nebulae. This makes it possible to travel conjecturally yet one stage farther back in time. Having already travelled back from earth to sun and from sun to nebulae, we can now complete our story by tracing the nebulae back to a mass of chaotic gas filling all space.

If the nebulae came into existence in some such way as this, we should expect them all to be of about the same size, weight, and intrinsic brightness. This is found very approximately to be the case. Two nebulae of the same shape often look very different in size and brightness, but the difference of appearance can usually be attributed almost entirely to their being at different distances from us.

If this is a general law, as at present it appears to be, then nebulae of any assigned shape may be treated as standard articles, just as Cepheid variables are, and their distances can be estimated from their apparent faintness (Plate LIII, facing p. 209). The faintest nebulae which can be photographed in the great 100-inch telescope at Mount Wilson prove to be so distant that their light takes 140 million years to reach us, and so are about a thousand times as distant as the farthest star in the Milky Way. Some two million nebulae lie within this distance.

Telescopically, these nebulae are of interest as forming very beautiful and interesting objects. Cosmogonically, they are of even greater interest as giving us a sort of cinematograph film shewing how we believe the sun and stars to have come into existence. Yet they have recently acquired an even stronger interest from the circumstance that they all appear to be running away from us—and this at perfectly terrific speeds.

We have already noticed how the motion of a star results in the lines of its spectrum being displaced—towards the red if the star is receding from us, and towards the violet if it is advancing towards us. Many of the lines in the spectra of the nebulae also are found to be displaced to abnormal positions, and this is most simply explained by supposing that the nebulae are themselves in motion.

Until recently, it was only possible to study the spectra of a few of the nearer nebulae, and these seemed to indicate that the nebulae were coming and going almost at random. Gradually it began to be noticed that the motions were not altogether chaotic; the approaching nebulae were mostly in one half of the sky, the receding nebulae in the other. All this could be explained if it were possible to suppose that the sun was advancing through space towards the former group of nebulae, and so of course receding from the latter group, at a speed of some hundreds of miles a second.

The cart-wheel rotation of the galaxy has now provided exactly the motion needed to justify this supposition. But the apparent motions of the nebulae prove to be something more than a mere reflection of the sun's motion through space. When the sun's motion is subtracted from the apparent motions of the nebulae, the nebulae are not brought to rest, and neither are they found to be moving at random, like the molecules of a gas. Instead of this we find that all the nebulae are receding from us with speeds which are almost, and possibly even quite, proportional to their distances.

In round numbers, each million light-years of distance is found to be associated with a speed of 100 miles a second. Nebulae which are a million light-years' distant from us recede with this speed, those at two million light-years' distance recede with double this speed, and so on. The largest speed of recession which has so far been observed is 15,000 miles a second—about a million times the speed of an express train. The nebula which holds this record is estimated to be at a distance of 135 million light-years, and so is very near to the limit of vision of the telescope.

When a shell bursts on a battlefield the fragments travel at different speeds, those which travel fastest also travelling farthest. At any particular moment after the explosion, each fragment will have covered a distance which is exactly proportional to its speed of motion. This is the same thing as saying that its speed is proportional to its distance from the point at which the shell burst. This is exactly the law of the receding nebulae, and makes it look as though, at some instant in the past, the universe had suddenly burst into fragments, our whole galactic system being one of the fragments—the particular one to which we are clinging.

There is however another way in which the motions of the nebulae can be explained. Imagine a number of straws floating down a river in company. If the river narrows at any particular spot, we shall notice the straws coming closer to one another, and where the river widens again, they will spread farther apart. When such a spreading apart occurs, an insect living on any one bit of floating straw will see all the other straws receding from it. And if the river has just passed through a very narrow bottleneck, their speed of recession will be exactly proportional to the distance, which again is the law of the nebulae.

Thus there are two possible explanations of the motions of the nebulae which look very similar, and yet there is a fundamental difference between them. When we compare the nebulae to the fragments of a burst shell, we imagine the nebulae to be moving *through* space. But when we compare them to straws floating in a river, the river must be space itself; the nebulae are not moving *through* space, but *with* space—they are straws shewing us in what way the currents of space are flowing, and the law that speed is proportional to distance suggests that space is expanding uniformly.

Probably the latter explanation is the best, because we now think that space is curved, and round, and finite in amount—rather like the surface of a balloon. Space is not to be compared to the air inside the balloon, but to the rubber which forms its surface. Thus we can travel on and on in space for ever, just as a fly could walk on for ever round the surface of this balloon. It would of course have to repeat its tracks, but it would never come to any obstacle that prevented its going on.

In the same way, we believe that if we tried to travel on for ever through space, we should never find anything to stop us, although sooner or later we should come back to our starting-point, as Drake did when he circumnavigated the globe. Needless to say there is no use in trying to circumnavigate space—for one thing life is too short. A ray of light might have a better chance, for it travels at 10 million miles a minute and is not limited to a lifetime of threescore years and ten. It was at one time thought that a sufficiently powerful telescope might let us look round space and see our own earth by light which, starting many millions of years ago, had travelled round the whole of space and finally come back to its starting-point. Naturally such an experience as this would give us a very direct and convincing proof of the curvature of space, but we no longer believe it to be possible to travel so far through space as this, even though we travel on the wings of light. Various astronomers have devised methods for estimating the size of the whole of space, and, however much they may differ from one another, they at least all agree that space is far too large for us to dream of seeing round it. The big telescope at Mount Wilson looks so far into space that we can see nebulae whose light started when our earth was inhabited by the weird animals we saw in our first chapter, and has been travelling over 140 million years to reach us. Yet it shews us only a tiny fraction of space, so small that it may perhaps bear the same relation to the whole of space as the Isle of Wight does to the surface of the earth.

Thus we see that not only is space almost inconceivably large, but it is continually becoming larger. It doubles its linear dimensions every 1300 million years or so, so that there is already eight times as much space as when the earliest radioactive rocks solidified, and perhaps more than a hundred times as much as when the earth was torn out of the sun. With every tick of the clock, its diameter increases by at least several hundreds of thousands of miles.

Possibly, however, we are more interested in matter than in mere empty space. Even in the tiny bit of space we can see there are some millions of nebulae, while in the part we cannot see there are probably millions of millions of nebulae, each containing thousands of millions of stars. Each nebula contains as many stars as there are grains of sand in a good handful, so that all the nebulae between them must contain about as many stars as there are grains of sand on all the seashores of the world. When we survey the vast universe as a whole, we see our sun reduced to a grain of sand, and our earth to a millionth part of a grain of sand—a tiny speck of dust circling round a grain of sand which is a million times bigger than itself, and yet is only of infinitesimal size in the universe as a whole. We may take pleasure in finding that the universe is such a very grand affair, but we cannot flatter ourselves that our mundane affairs play any large part in it.

Such is the universe of our travels. If we have not been able to construct a complete cinematograph film, at least we have seen a series of pictures on which something of its past history has been sketched. We first saw a primaeval universe which consisted merely of a mass of chaotic gas. As we watched, we saw this gradually condensing into nebulae. It seems very probable—although I do not think this has been strictly proved as yet—that such a condensation of chaotic gas into nebulae would of itself start space expanding. At any rate, for this or for some other reason, space itself began to expand, which means that even while the nebulae are forming, as well as for ever after, they must move steadily farther away from one another.

During all this time, the nebulae are changing their shapes in the way we have noticed, until finally they end by breaking up into stars. One particular nebula was the birth-place of our own familiar friends, Sirius, Aldebaran, Arcturus, and so on, as well as a far smaller and less brilliant object—our own sun. For millions of years, these and millions of other stars move blindly past one another, until finally we see our sun wander into the danger zone of a bigger star, and a cataclysm results out of which the planets are born—our earth among others. At first it is simply a ball of hot gas—as the sun is now, but much smaller. In time it cools down, liquefies and finally forms a solid surface; we see steam condensing into water, and forming seas and rivers. Then—greatest mystery of all—life appears. It is very humble at first but gradually increases in complexity until finally, only a few minutes back on the astronomical clock, man emerges, and starts gradually and slowly to climb the long steep ladder of civilisation. Yet only within the last few ticks of this clock has he concerned himself with the meaning of the nightly pageant of the sky. Then Egyptians, Chinese, Babylonians and Greeks began in turn to wonder what it all meant. Only one tick ago the telescope was invented and gave us the means of finding out. Within that one tick almost all I have told you has been discovered, and many thousands of times as much besides. And with our knowledge of the skies increasing at its present rate, who shall say what strange surprises the next tick of the clock may have in store for us?

INDEX

CAMBRIDGE: PRINTED BY WALTER LEWIS, M.A., AT THE UNIVERSITY PRESS

Ingram Content Group UK Ltd.
Milton Keynes UK
UKHW020649240723
425668UK00005B/344